Environmental
Design Evaluation

Environmental Design Evaluation

Arnold Friedmann
University of Massachusetts
Amherst, Massachusetts

Craig Zimring
Georgia Institute of Technology
Atlanta, Georgia

Ervin Zube
University of Arizona
Tucson, Arizona

PLENUM PRESS • NEW YORK AND LONDON

Library of Congress Cataloging in Publication Data

Friedmann, Arnold.

Environmental design evaluation.

Bibliography: p.
Includes index.
1. Architecture – Environmental aspects – United States. I. Zimring, Craig, joint
author. II. Zube, Ervin, joint author. III. Title.
NA2542.35.F75   729   78-24252
ISBN 0-306-40092-8

Graphics and Figures by Alyce Kaprow

©1978 Plenum Press, New York
A Division of Plenum Publishing Corporation
227 West 17th Street, New York, N.Y. 10011

Printed in the United States of America

Preface

As the nature of the field of environment–behavior relations is interdisciplinary, the collaboration of three persons of diverse professional backgrounds in writing this book is therefore not surprising. This collaboration started in 1972 with the offering of a graduate seminar "Environment, Behavior, and Design Evaluation" at the University of Massachusetts. Several research projects dealing with design evaluation which have been conducted at the University are also included as case studies in this book (Chapter III): the ELEMR study and the Visitor Center study.

Two of the authors have worked as part of the instructional team in the seminar, and all of the authors have participated in varying degrees in the ELEMR Project. The authors' backgrounds in design, psychology, and landscape architecture suggest, by example, that professionals with diverse backgrounds but a common interest in environment–behavior problems can indeed learn to communicate and to collaborate.

Since design evaluation is a new field and very little specific literature on the subject exists to date, we hope this book fills a current need. In developing plans for the book, it became obvious to us that a theoretical orientation toward the subject would be of little direct help to design practitioners and students. Hence we arrived at the current format of the book, which also provides an assessment, albeit indirect, of the state of the art. We hope the book will serve as a text for students, as a reference and guide for design professionals and behavioral scientists, and as a work of general interest to all those involved in this field. We have attempted to abstract sometimes lengthy case studies while retaining important distinctions of study purpose and methods, and we have tried to avoid the use of jargon in these presentations.

The book consists of five chapters. Chapter I introduces the subject of design evaluation and discussed major conceptual issues, drawing upon the case studies which follow as a means of illustrating the issues. The next three chapters consist of case studies dealing with interior spaces, buildings, and outdoor spaces, respectively. Each of these chapters has a brief introduction which directs the reader to important distinctions among the case studies. The final chapter summarizes our thoughts on environmental design evaluation and discusses methods, techniques, and various special considerations for evaluative studies.

For many readers it might be best to read Chapter I and V first in order to gain an overview of the subject. Depending upon specific interest, the reader can then select the most relevant case studies or can proceed chronologically with all of the studies. Each case study heading provides the title, names of researchers, methods used, description of the type of project, and information source. We also refer the reader to a rather extensive bibliography which contains both specific design evaluation and methodological references and references on the general subject of environment–behavior relations.

We wish to express our thanks and appreciation to the many researchers who permitted us to use their reports as case studies for inclusion in this book. We also wish to express appreciation to the many colleagues who have advised us on both format and individual cases, and who have read the manuscript and provided valuable criticism. Foremost among those are Joseph Crystal, Joanne Green, Geoffrey Hayward, Min Kantrowitz, R. Christopher Knight, Jan Reizenstein, and William Weitzer. We extend special thanks to Stanley Moss who was a strong supporter of the ideas behind this book and our colleague in the graduate seminar over the last 5 years.

Every effort was made when abstracting the original studies not to alter the meaning or leave out key points. We accept responsibility for any errors of fact or erroneous interpretations and extend an apology to those whose work we may have misinterpreted. We are also grateful to those who have allowed us to reproduce their photographs. Original artwork for many of the charts, graphs, and maps was not available to us so they have been redrawn for inclusion in this book. We thank the authors for permission to use these, and extend our gratitude to Alyce Kaprow for her excellent work in preparing them.

Amherst, Massachusetts   Arnold Friedmann
Atlanta, Georgia   Craig Zimring
Tucson, Arizona   Ervin Zube

Contents

I

Introduction
A Structure–Process Approach to Environmental Design Evaluation

Frank Lloyd Wright once defined design as "art with a purpose." Indeed, the history of design supports Wright's words. Gothic cathedrals were intended to instill an awe of God in worshippers; Bauhaus-inspired structures were intended to function as efficient machines; modern low-income high-rises are designed to lift their tenants into the middle class. All of these diverse purposes have a common element—they are oriented toward people.

In the past few years, Wright's words have begun to take on special significance. Terms like environmental psychology and behavioral architecture have been coined to describe the growing belief in two interrelated ideas: (1) The designed environment affects human experience in direct and important ways. It does not *determine* experience, yet in combination with social influences, designed environments can support satisfaction, happiness, and effectiveness. (2) Despite their potential, designed environments often do not "work" with respect to their impact on human experience. They are awkward, even destructive, rather than being supportive of personal competence and growth.

The Pruitt-Igoe apartments in St. Louis are a famous and poignant example. Although hailed as a solution to the slums when it first opened in the 1950s, the design did not reflect an understanding of existing community patterns. As a reaction against crowded, deteriorating housing, the designers razed existing three-story tenements and replaced them with high-rise buildings on a multiblock site. While the low-rise tenements had permitted surveillance of the street and the monitoring of children, the high-rise buildings separated tenants from their normal street life, encouraging crime and destroying a sense of community. By

1

failing to understand the users' needs, the designers had produced a plan which most emphatically did not work for its intended users. The apartments became vacant; finally, several years ago the city of St. Louis dynamited them on national television.

How are we to make our designed environments work better? We must concern ourselves with all aspects of the building design process: fine-tuning existing structures, improving new designs, improving minimum building standards and life-safety codes, clarifying user needs and preferences, and improving public housing programs. Clearly, we must retain and improve ideas that succeed and abandon ideas that fail. In order to do this, we must carefully analyze and understand our designs. In this book we will discuss several methods of analysis, as we discuss the process of environmental design evaluation. Environmental design evaluation will be defined as "an appraisal of the degree to which a designed setting satisfies and supports explicit and implicit human needs and values." — what about the satisfaction of the designers intent then G. + O.

WHAT DISTINGUISHES ENVIRONMENTAL DESIGN EVALUATION FROM OTHER APPRAISALS?

We make appraisals of the designed environment every day. We patronize pleasing restaurants and avoid unpleasant ones; we go to a park we like and don't go to one we dislike; we walk a few extra steps to pass through a grand entrance to a building rather than taking a less imposing shortcut. These are evaluative judgments and as such are not so different from the ones which will be discussed in this book.

If we are to improve the practice of design, however, our appraisals must be careful and systematic. Appraisals are needed that represent each subgroup of users and that examine each important design element. Individual judgments by the designer and the lay person must not be abandoned, but must be augmented by more complete and rigorous techniques. These methods are discussed in the studies in the following chapters.

Yet, except for rigor and an orientation toward users, it is often hard to discern a common theme in the evaluation studies in this book. Rather, they show considerable diversity. For example, these studies range from a several-hundred-thousand-dollar longitudinal evaluation of a state school for the developmentally disabled to several evaluations performed by graduate students as class projects.

However, as diverse as the study methods are, they have a common origin: a firm foundation in the social sciences. Evaluation studies often gather behavioral information, use methods drawn from psychology or sociology, and, in fact, are often performed by social scientists. Yet because of the real-world nature of the settings, evaluation studies tend to be different from studies in experimental social science in several different ways: whereas social science strives to *control* extraneous factors, evaluation often *describes* those factors; whereas social science is most concerned with discovering *causes* for behavior, evaluation looks at *influences* on behavior; whereas social science strives to use *causal* statistical models, evaluation uses *correlation* models; whereas social science aims to reduce the number of factors, evaluation often examines complex systems. In short, evaluation does not attempt to reduce complexity, but rather attempts to conceptualize it. Indeed, in this book we will present an informal systems approach to the evaluation of designed environments.

We will briefly describe a method of organizing evaluation studies that has been effective in guiding several evaluations. Chapter I describes a general approach to environmental design evaluation, including (1) a structure (the information needed) and (2) a process (the evaluation process itself), while Chapters II through IV examine specific evaluation studies by setting. Chapter II surveys several successful case studies which evaluate building interiors; Chapter III deals with case studies which consider buildings-as-complete-systems; and Chapter IV evaluates several cases which consider building sites and open spaces. Chapter V summarizes environmental design evaluation by raising some critical ethical concerns, and by outlining available methods.

THE NEED FOR EVALUATION

The need for rigorous environmental design evaluation is gaining recognition in several professional groups of behavioral scientists, designers, and legislators. Indeed, while behavioral scientists have long studied the social environment as it affects human behavior, they have tended to ignore the role of the physical environment. The recent surge of interest in the field of environment–behavior research, however, has spotlighted the role of the designed environment. Settings as diverse as psychiatric institutions, schools for the retarded, college dormitories, offices, homes, and undersea laboratories have all been scrutinized by social scientists. Also, the physical environment has been increasingly considered in investigations of crowding, personal space, privacy and other behavioral processes.

William H. Whyte (1972a) commented recently on the failure of

designers to observe the consequences of their actions and to systematically learn from past experiences. He called for altering this condition and for including design evaluation as an important part of landscape architecture curricula. The American Society of Landscape Architects recently endorsed launching a major research effort directed at design and planning evaluation (ASLA, 1974). The recommendation was ranked as one of the highest-priority issues facing the profession. The discussion in support of the recommendation stated:

> The systematic analysis and evaluation of completed works (i.e. design evaluation) provides the greatest potential for obtaining the kinds of data and knowledge essential to improving professional performance. Systematic approaches should provide the basis for comparative as well as case studies. The findings would be of value for the continual iterative up-dating of educational programs as well as for the prediction of impact of design-planning decisions by providing a more substantive information base on which to make such decisions. (p. 6)

Indeed, a number of authors have discussed the failure of design professionals to consider evaluation as an essential ingredient in the design process (Lang *et al.*, 1974; Rapoport, 1969, 1973; Whyte, 1972a; Zeisel, 1973). The design process as followed in most professional practices today usually terminates with the construction of the project. What evaluation does occur is inclined to be sporadic, limited in scope, and idiosyncratic in approach. Data from these evaluations tend to be noncumulative and noncomparable; they do not really alter the design process. This is unfortunate since evaluation has the potential to be an exceedingly effective educational device within the profession. The systematic description, analysis, and evaluation of completed works can provide valuable information not only to educate aspiring professionals in design schools, but also to update the established practitioner.

Design evaluation needs to be incorporated into the considerable number of publicly financed programs which are directly related to environmental design. For example, in 1970 the federal government spent approximately $47 million in aid to state and local governments and for the purchase, development, and operation of city recreation and park facilities, and spent an estimated $150 million in 1976, the last year for which figures are available (Council on Environmental Quality, 1977). Federal housing support has ranged from the construction of public housing to home mortgage insurance programs, loan guarantees for new towns, and regulations for institutions for the mentally retarded. Some of these programs, such as the Federal Housing Administration, have been in existence since the 1930s. Although the environments these programs have created affect the lives of millions of citizens, there have been virtually no attempts to establish a systematic, comprehensive approach to the evaluation of the resultant environments, either by the funding agencies or the design professions. Since these environments, as measured by their capacity to support and satisfy human needs and values, are one of the best indicators of the success or failure of the federal programs themselves, evaluations of the environments resulting from these programs would be an important addition to the recent emphasis on management-focused analyses of federal programs.

There are indications that public policymakers are indeed beginning to respond to the need for evaluation. The National Environmental Policy Act of 1969 (NEPA) set the stage by requiring both the preparation of environmental impact statements and public disclosure of the potential impacts of major projects. This act is in fact an explicit statement of values which underlie the need for design evaluation. The public disclosure aspects of NEPA both encourage citizen involvement in the decision process and tend to force designers to state more explicitly their assumptions and values. In the interpretation of this act the courts are adopting a broad definition of environment, one that includes social, aesthetic, and economic as well as physical and biological parameters (Baker, 1973). The definition and the provisions of the act encompass the designed as well as the natural environment. The act also explicitly calls for the use of:

> a systematic, interdisciplinary approach which will insure the integrated use of the natural and social sciences and the environmental design arts in planning and in decision making which may have an impact on man's environment.

In this statement, NEPA calls for the *prediction* of the environmental consequences of proposed developments. Prediction presumes, however, the existence of some substantive empirical and/or theoretical base, and yet, no such base exists. The evaluation studies in this volume and elsewhere provide a promising beginning.

In summary, there are several related reasons for the development of design evaluation programs:

1. To extend our understanding of human behavior by further documenting the transactions of people and the built environment.
2. To extend the design process to include evaluation and the development of a feedback mechanism for the inclusion of evaluative data in the making of design decisions, both for fine-tuning existing environments and creating new ones.
3. To provide an important body of data for use in the education of

future design professionals and for use in continuing education programs.

4. To obtain the kinds of data required for the analysis of the efficacy of public policies and programs which support and constrain the design of a range of environmental settings.

5. To begin to develop a capability for the prediction of user satisfaction and environmental fit for environmental impact assessment in its broadest definition.

THE STRUCTURE–PROCESS APPROACH

The structure–process approach has two requirements for guiding evaluations: a mental structure for organizing information and an orderly process for actually completing the research. While these two components can be discussed separately, they are as interrelated as—to paraphrase W.B. Yeats—the dancer and the dance. The structure presented here is a five-part conceptual scheme which helps to organize the information that must be considered in an evaluation; it is a static model for organizing information. The process presented is a multistep procedure—an action plan—which helps to organize an actual evaluation study. Both the structure and the process interrelate and alter the other: the way in which we think about a setting suggests methods of evaluating that setting, while the ways that we evaluate a setting change our thinking about it.

Structure: Information to Consider

While evaluation can be conceptualized in many ways, it is useful to have a general scheme which helps in organizing our knowledge of the situation, establishing models, and focusing our conclusions. Such a scheme must consider at least four factors of the appraisal process: (1) the *setting*—the social and physical attributes of the designed project being evaluated; (2) the *users*—the background, needs, and behavior of the people who are involved with the setting, such as tenants, customers, maintenance workers, and managers; (3) the *proximate environmental context*—the ambient qualities, land-use characteristics, and neighborhood qualities that surround the setting; and (4) the *design activity*—the activities by designers, regulatory agencies, clients, and users which resulted in the final design of the setting. Moreover, these four factors all exist in a *social–historical context*, a larger society in which one must consider larger-scale social, economic, and policy issues, such as social

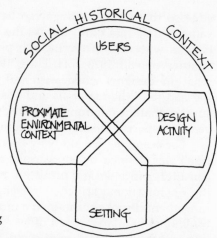

FIGURE 1. The structure for organizing
information in an evaluation.

mores, unemployment levels, and demographic profiles. (These factors
are illustrated in Figure 1.)

Consider the case study of National Park Service visitor interpretive
centers by Zube *et al.* (1976)(see Chapter III). The centers themselves
were the setting; visitors, staff, maintenance workers, and managers
were the users; the parks in which they were situated and the local
weather conditions formed the proximate environmental context; and in
some cases, exhibit designers were involved, while in other cases they
were not, forming a complex design activity. The social–historical context was the society in which the setting, the users, the proximate environmental context, and the design activity existed. For example, more
positive attitudes toward the outdoors produced many more visitors to
National Parks, bringing problems of crime and overcrowding.

Although these factors are highly interrelated, for simplicity we
discuss them separately in the following section. It should be kept in
mind, however, that these factors are discussed in separate sections
simply as a heuristic tool; this division is not meant to imply they are
independent of one another. Each element is discussed in general,
examples are drawn from an evaluation study presented in later chapters, then specific concerns are addressed.

The Setting

Many postconstruction evaluations have focused on user satisfaction, and understanding satisfaction is certainly a critical goal of evaluation. But if one of the purposes of design evaluation is to provide knowl-

edge so that future settings can be improved, one must question the value of data that only address the users' values, perceptions, and behaviors without considering the physical environment to which they relate (Wohlwill, 1973; E.H. Zube, 1974). The physical setting is, after all, what the designer works with and produces. Important issues include the nature of the materials and spatial relationships, the limitations or support that the environment affords, and the relationships between the various aspects of users' satisfaction and the physical environment.

For example, in a study by Brookes (1972)(see Chapter II), the "setting" was defined by the various physical and social elements which constitute office design: furniture, materials, noise, and light, as well as the organizational structure which dictated office sizes and office arrangement. Brookes examined a move by a large corporate office from a traditional to a landscaped office. Prior to the move, the bright, noisy offices were furnished with a sea of traditional gray and green desks arranged in large rectilinear patterns on tile floors, with only a limited access to windows. The secretaries shared large open spaces, the supervisors had cubicles defined by head-high partitions, and some executives had private offices. After the change, most desks were separated by movable screens, the office was carpeted in strong colors, new furniture was purchased, and bright colors, graphics, and plants were added.

These physical and social elements of a "setting" were expanded to include structural building elements in a study of English housing complexes by Reynolds and Nicholson (1972)(see Chapter IV). The characteristics of the four building forms—row houses, slab blocks of 14 to 22 stories, square point blocks of 11 to 21 stories with access to balconies, and square blocks of 3 to 14 stories—affected other qualities, such as density and provision for outdoor space, which strongly affected tenant satisfaction.

In summary, a description of the setting should include, where relevant:

1. Organizational goals and needs, such as: Which factors such as user satisfaction, productivity, or sales are most highly valued and by whom?
2. Organizational functioning, such as: Which groups affect each other? Which groups need to communicate? Which groups, such as sales staff in a store, or direct-care staff in an institution, most directly affect the values and goals mentioned above?
3. Relevant materials, structural elements, spaces, and design solutions;
4. Important ambient qualities such as noise, light, and temperature;
5. Elements which have symbolic values for the various user

groups, such as differential office sizes in business, bars in the windows in institutions, or signage and graphics in schools;

6. Provisions made for access by groups with special needs, such as ramps for users in wheelchairs or Braille signs for the blind;

7. Condition of the setting and of temporary elements such as the quality of maintenance or changes and decorations provided by the users.

The Users

The ultimate test of the success of a designed setting is its ability to satisfy and support explicit and implicit human needs and values; that is, to provide a physical and social milieu within which individuals' and groups' lifestyle aspirations are reinforced and values are recognized. If we are to evaluate such a correspondence between the milieu and the needs of users, however, we must understand who the users are. This is not a trivial concern. Many of the users may be quite removed from the traditional designer–client relationship. For example, in evaluating an office building, the clients may be the corporate board of directors, yet the users may include such diverse groups as executives, managers, sales staff, secretaries, customers, maintenance staff, and even passersby who are affected by the building. The major problems in an evaluation are often defining these user groups, understanding the characteristics which describe them (such as age, income, and organizational position), and understanding the distinct needs of each group. The age of users may suggest whether benches, play equipment, or toddler areas are important; ethnic identity may help to understand whether the locus of family activity will be in the family room, dining room, or kitchen; the income level of a community may suggest the amenities that will be most used.

In addition, the designer must understand what the residents really *do*. People may say that they like something, but do they use it? Are the existing stereotypes for various racial and ethnic groups accurate reflections of people's needs and activities or merely popular misconceptions?

There are several ways to find out what people do. Of these techniques, interviews and questionnaires share the problem that respondents often tailor their responses to make them "socially acceptable." Observations of overt behaviors are often less susceptible to this pitfall, especially if behavior-recording methods are used which do not affect the observed behavior. These methods include the use of records and other "behavior traces" that are discussed in the section dealing with methods.

For example, in a study of an urban park, Nager and Wentworth (1976)(see Chapter IV) used a justifiably broad definition of user: In addition to questioning park users, they polled employees in nearby businesses who did not use the park. In this sense "users" included people who did not actually visit the park. This was an eminently reasonable decision since this procedure helped to understand why some people did not use the park. A park is an important public amenity which aids in establishing an image of the city even for nonusers. Indeed, nonusers saw the park as much more dangerous and much less relaxing than did users.

The overt behavior of users was a prime focus of the ELEMR Project. (see Chapter III) This 3-year project studied the impacts on residents and staff of providing semiprivate resident living spaces, carpeting, and other renovations at an institution for the developmentally disabled. Prior to the renovation each resident had slept in a large open ward with 25 other residents. After the renovations each resident had a more private arrangement, including a semiprivate module with a chair, a dresser, and a bed, or a small bedroom shared with one to three other people. The ward users consisted primarily of residents and the direct-care staff who dealt with them every day. The residents often lacked communication skills, and the direct-care staff were firmly entrenched in an organization in which they had little power. Since these characteristics made it difficult to get unbiased verbal response from either group, coded observation of the residents' and staffs' overt behaviors were used as the evaluative factors, rather than using verbal or written responses.

In summary, a description of the users should include:

1. Perspectives, preferences, needs and attitudes, such as need for privacy, spatial utilization, aesthetic preferences, cultural values, images of setting and context, and environmental satisfaction;
2. Behavior in terms of individual and group activity patterns, social behavior, and behavior variation over time and space;
3. Individual characteristics, such as age, sex, income, education, and ethnic or cultural background.

The Proximate Environmental Context

We have suggested that it is necessary to understand the designed setting and the users who interact with it. However, since each designed setting exists within an immediate physical and social context, this context may be as pervasive as poor air coming from a nearby steel plant, or may be as specialized as hilly terrain which limits movement only for some groups, such as the elderly. Indeed, each setting exists in a neighborhood and is affected by local conditions of climate, air quality, water quality, transportation, cultural facilities and programs, and safety.

In a study of a campus open space, Cohen *et al.* (1976) (see Chapter IV) found that the location of the space—its proximate context—helped describe both existing use patterns and future needs. The space was bounded by the Student Union and university library and by several classroom buildings. Dormitories were located to the northeast, southeast, and southwest. The central location of the space helped predict that most students used the space as a traffic corridor. The central location also suggested that the space had potential importance as a recreation space and reference point. Hence, issues of context were central in the evaluation.

In another example in Chapter IV, Rutledge (1975) evaluated a design in an intensely urban context in a study of the First National Bank Plaza in downtown Chicago. The plaza was bounded on three sides by narrow, traffic-congested streets lined with tall office buildings. Once again, context was critical. The center city urban context helped explain why most users frequented the plaza during the lunch hour and why the majority were white, male, white-collar workers.

Finally, in their study of National Park Service visitor interpretive centers, Zube *et al.* (1976)(see Chapter III) adopted a more physical definition of proximate environmental context. Because they are situated in national parks, interpretive centers often are subjected to extreme climatic conditions. For example, since Olympic-Hoh in Washington receives over 200 inches of rain a year, roof leakage is a serious problem. Also, a park becomes a "neighborhood" for an interpretive center, providing unique problems of access in case of fire, exposure to wildlife, and so on. Hence, physical environment and topography are important contextual factors.

In sum, an evaluation should consider several factors in the proximate environmental context such as:

1. Environmental and ambient characteristics, such as noise, air quality, climate, drainage and topography, vegetation/soils, and aesthetics;
2. Land use, such as quality and type of neighborhood, density, mix of uses;
3. Supportive facilities and programs, such as accessibility, transportation, cultural facilities, and safety.

The Design Process

Perhaps the single most ignored aspect of environmental design evaluation to date is that of the design activity, of gaining insight into the process which produced the setting. The designer is only one of the actors in a complex process (Cooper and Hackett, 1968). Other "actors" who can play roles in this process are the client, who may not be the user, the users, the financier, boards and committees, and public officials and agency representatives responsible for the administration of different public projects and programs. Each individual brings to the design activity or process a set of values, preferences, attitudes, and limitations which are in part conditioned by his or her role and which have to be accommodated in some way in the production of a design setting. Many decisions are also made by these actors before the designer becomes involved. Location may have been determined, funding programs identified, and major design program elements decided upon (Cooper, 1968). The extent to which there is agreement on values, preferences, and attitudes among these actors and the designer and users can be an important issue in the creation of settings which satisfy and support the users' needs and values. For example, the problems which accompanied the low-income Pruitt-Igoe apartments came in part from a lack of congruence between the perceptions of designers and some user groups

In addition to the cast of actors in the process there is another important factor which impinges on the design activity. It consists of the numerous directives, limitations, and criteria which influence the form of the designed environment. These include municipal zoning ordinances and subdivision bylaws, legislation, and policies and administrative guidelines for federal housing and open-space programs. These elements constitute an important evaluative factor, a set of constraints on the design process that is effectively outside the control and influence of the designer. They may be susceptible to modification over time, however, if empirical data are available to show shortcomings and to illustrate alternative policies and guidelines which could lead to more satisfactory and supportive environments.

Another aspect of the design activity that is easily overlooked is that of environmental changes which are made by the users after the completion of the project. Such changes may represent efforts to induce flexibility into a setting, to personalize and individualize an environment. Such changes may also signify post hoc efforts to compensate for incongruent designer–user perceptions and values.

For example, a major goal of Zeisel's and Griffin's (1975) evaluation of the Charlesview Housing apartment complex (see Chapter III) was to identify the designers' intentions and expectations and then compare those intentions to the completed design. They discovered, for example, that the architects intended a clustered site plan to convey a "hierarchy of spaces," yet residents did not understand this plan. Residents

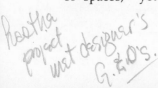

knew other people from their immediate area no better than they did residents from other areas; the intended design was only a partial success.

In addition, Zeisel and Griffin gained an understanding of the sequence of events comprising the design process. Although the urban site was originally intended for high-rise, high-rent housing, a public outcry resulted in the construction of low-rise, low-and-moderate-income housing under the sponsorship of an ecumenical nonprofit corporation, who in turn hired the architects. Except for zoning controls placed on the parcel by the redevelopment authority, the architects were not given any special instructions. They developed a series of alternative plans and arrived at the final design by synthesizing their clients' reactions to these various possibilities. Unlike other cases (Cooper, 1968), the architects had considerable influence over the final form of the project.

In a second case, the Research and Design Institute of Providence, R.I. (REDE, 1974) had a unique opportunity to evaluate their own design for the renovations of Butterfield Hall, a college dormitory (see Chapter II). REDE recorded the design process as it occurred. The designers used an open design process in which they assessed the needs of Butterfield Hall residents through open meetings and interviews. The resulting designs included a variety of room arrangements, an activities corridor, a flexible furnishings system, and a geodesic roof structure. The latter was to provide space for recreation such as ball playing, which had taken place in the hallways before renovation. The open design process resulted in a generally well-received design, with a notable exception. Cost cutting had eliminated the roof structure, and the students who had participated in the design process felt especially cheated by the final outcome.

Finally, Reizenstein et al. (1976)(see Chapter II) performed all the steps in the design cycle for renovations at the Cambridge Hospital Social Services offices: predesign programming, design, and postoccupancy evaluation. They interviewed and observed the staff members to document their activities and physical requirements and, in turn, translated their findings into design specifications. The predesign programming interviews helped identify important evaluative factors for the final design. These included use of the offices for meetings or to see clients, the image of the offices for other hospital personnel, and adequate size for private offices. The evaluators' programming interviews led to the final evaluation: responses to questionnaires administered after the completion of the renovation revealed that there was moderate to significant improvement in the staff's feelings about these issues.

In summary, a description of the design process should include several evaluative factors:

1. Roles of the participants including the decisions made by designers, clients, financier, user, and public officials;
2. Values, preferences, and assumptions of the various actors, both about user behavior and about different aspects of the setting;
3. Constraints that helped form the setting, such as budget or codes and ordinances;
4. Postconstruction modifications by users, managers, or designers. *how users react to a given design*

The Social–Historical Context

We have suggested that designed environments can be conceptualized in terms of four information factors. Yet, designed environments are not independent isolated projects; rather, they exist in a complex society which changes over time. Many times a project may be a success or a failure when it is judged as a single isolated entity, yet it takes on a new light when viewed in a larger context or at a different time. For instance, as of this writing there is heated controversy surrounding Richard Meier's Bronx Developmental Center. The building is an aesthetic masterpiece; it has won a plethora of major architectural awards. Yet the philosophy that was popular during the formulation of the design program for the Center is now obsolete: the building is still institutional, it separates residents from society rather than integrating them. The building is an aesthetic success, yet it is a failure for its users.

Indeed, viewing these four factors from a larger context informs evaluations. A park may be a beautiful aesthetic statement yet ignore the growing proportion of elderly people in the population and fail to provide the necessary amenities. A welfare office may provide a strong aesthetic statement through minimal design, yet its design may alienate unsophisticated users. While a ponderous social analysis is not required in these cases, it should be recognized that designed environments are not passive, neutral physical entities; rather, these environments are complicated systems that take on new meanings as society changes, as political currents shift, as the economy improves or worsens, as philosophies, styles, and mores evolve.

For the designer, the social–historical context is traditionally incorporated into the design process. In part the larger issues are indirectly addressed in the design program given the designer. Designs stand for 30 years or more; really good designs are created when the designer can go beyond the program and create projects which anticipate social, eco-

design in time

flexibility valued n trade-off

"anticipatory" nature of design = good design
anticipatory, flexible, timeless sol <—> practicality

nomic, and political change. The Bronx Developmental Center evidently did not do so.

The Reizenstein and McBride case discussed in Chapter III provides a clear example of how a larger social perspective may inform an evaluation. The evaluators studied a small village for developmentally disabled people which was based on the "normalization" principle. This principle dictated that the residents be given life patterns that resembled everyday life—one of these being a separation of work and leisure. The village was aesthetically pleasing, yet its workshop combined recreational and work functions. Since this combination was confusing and did not meet the normalization goal of separation of work and leisure, the village received a lowered evaluation.

In sum, the four informational factors described above need to be viewed in terms of a larger social–historical context. A description of this context might include:

1. Social and political trends which might affect the setting, such as the economic climate, treatment philosophy, or social attitudes;
2. Historical changes in these trends, both in terms of the past and of the projected future.

The setting, users, proximate context, and design process must be seen in terms of the larger society, the social–historical context. Within each of these, we have identified a number of factors which should be considered individually in a careful evaluation. These are illustrated in Table 1. It is our contention that although the large number of evaluative factors may appear disheartening, each of the five should be carefully considered in an evaluation. While each factor can be explicitly considered during the formulation of an evaluation (and such consideration is an important aid in designing the evaluation), for reasons of time, interest, or resources some factors will be explored in greater depth than others. In the next section we will examine the issues involved in choosing the relevant factors and in understanding their interrelationships.

 ## Process: Steps to Follow in Designing an Evaluation

Designing an evaluation study is not so different from designing an environment. Research design, like environmental design, is an interrelated series of decisions, and although some decisions precede others in time, each decision affects every other. The intended use of the evaluation helps to define the problem; the definition of the problem helps to establish methods and suggest methods; and the methods determine the form of the information and influence its eventual uses.

In the next two sections we will discuss some important decisions

TABLE 1
Evaluative Factors

The Setting	*The Users (cont.)*	*The Proximate Environmental Context (cont.)*

Organizational goals and needs
  Communication
  Values
	Income
	Ethnicity
		Density
		Distribution/location
		Area

Organizational
  Functioning
  Who affects whom
  Management style
	Group characteristics:
	  Lifestyle
	  Stage in life cycle
	  Socioeconomic status
	  Values
		Supportive facilities and
		  programs
		  Accessibility/transportation
		  Cultural facilities
		  Safety

Materials
  Form, color
  Style
  Texture
	Perceptions,
	  preferences, and
	  attitudes
	  Privacy
	  Spatial utilization
	  Aesthetics
		The Design Activity

Ambient qualities
  Noise
  Microclimate
  Air
  Light
  Natural or manmade
  character
	Image of setting and
	  context
	  Environmental
	  satisfaction
		Participants
		  Roles: designer,
		  client, financier,
		  user, public officials
		Values, preferences,
		  attitudes,
		  assumptions as to
		  user behavior and
		  setting

Symbolic elements
  Institutional features
  Status symbols
	Behavior
	  Individual and group
	  activity patterns
	  Social interaction
	Spatial variation
	  Temporal variation
		Constraints
		  Budget
		  Public Policy
		  Codes and ordinances

Provisions for
  handicapped
  Ramps
  Braille signs
	The Proximate Environmental Context	Postconstruction
		  modification
		  By users
		  By manager
		  By designer

Transitory elements
  Upkeep
  Decorations by users
  or others
	Environmental
	  characteristics
	  Air quality
	  Noise
	  Climate
	  Drainage and
	  topography
	  Vegetation/soils
	  Aesthetics
		The Social–Historical Context

The Users		Social trends
		  Economic
		  Treatment philosophy
		  Social

Individual
  characteristics:
  Age
  Sex
  Education
	Land use
	  Type of mix
		Historical changes
		  In above trends

that must be made when designing an evaluation. First, the preevaluation decisions: Who is responsible for initiating and compiling environmental design evaluation? Who should participate in the design evaluation team? What is to be evaluated? Second, the evaluation process itself, including: What is the focal problem? What is the larger system? How do you define methods? How do you analyze data? How do you put the data into the design cycle?

Who is responsible for initiating and compiling evaluations? To date, many evaluation studies have been initiated by social scientists as a part of the growing field of environment-behavior research. Indeed, evaluation helps to address important social science questions: What situations are chosen by different types of people? What elements in the environment are most important to people? How can designed environments be used to facilitate happiness, productivity, therapeutic progress? Despite the importance of such questions, there is a serious shortcoming in evaluation research to date: little aid has been offered to designers. More information is critically needed that is directly pertinent to the task of shaping and manipulating environments (Appleyard, 1973).

Social-science-oriented research must go on. We suggest, however, that the lack of design-oriented information must be addressed by the design profession itself. Indeed, the responsibility is theirs to advance the delivery of design services and to incorporate those elements of the social and behavioral sciences which contribute to that end. Within the profession, there are two logical foci of evaluation activities: the design schools and the active practitioners.

The design schools can serve as active centers for two activities: (1) for teaching design evaluation as part of the normal design curriculum and on an in-service basis; and (2) for collecting and disseminating evaluation research data. Whyte has suggested that the education of design students in evaluation has two aspects:

1. Teaching the student how to look at the environment and to find links between form and behavior;
2. Teaching the student what the people who specialize in looking (social scientists) can do for the profession and what they can't (Whyte, 1972a).

In addition, design schools can initiate student and faculty research programs, perhaps in collaboration with members of the psychology or sociology departments. Many of the studies in this book have resulted from such initiatives. Moreover, funding for projects may be obtained from a variety of sources. Funding for the studies in this book were internally generated, were provided by sponsors of the project being evaluated, or came from outside sources such as HEW or HUD. Once the schools initiate research programs, it is a logical extension for the schools to establish data banks of evaluation information. There is a growing precedent in other disciplines for information to be shared regionally; this is a useful model for evaluation information as well. Such a data bank would be accessible to students, educators, and practitioners alike.

The second logical focus of evaluation activity—the design practitioners—is in some ways more problematic. The motivation for evaluation is simple: better designs result. Yet to actually integrate evaluation into the design process two elements must be present: (1) designers must use evaluation information as they formulate their designs, and (2) designers must evaluate those designs when they are completed.

To use evaluation information in their work, designers must have ready access to relevant information in the right format. We have suggested that the schools could create databanks of information. Perhaps the professional organizations such as AIA, ASID, or ASLA can serve to bridge the schools and the practice. Second, if designers are to evaluate their designs, they must be properly trained and properly rewarded to do so. For example, we have suggested that the design schools should offer both regular and in-service training in design evaluation.

The issue of adequate reward is more difficult. Often even a successful practice does not allow for elaborate evaluations—except for the largest firms. Clearly, fees have to be earmarked specifically for in-depth evaluation. For example, in cases of large government clients such as HUD, HEW, or GSA, legislation may be initiated which dictates an additional 5% fee for evaluation. Or large public-sector clients may be persuaded that their subsequent schools, offices, apartments will be more successful due to evaluation, and hence might contribute a similar fee. Yet even if such add-on funds are not forthcoming, evaluations need not be abandoned for lack of funds. Indeed, several of the evaluations in this book were generated using in-house funds and required only a few person-days.

Who should participate on the design evaluation team? Evaluation provides important information for both social scientists and designers and contains elements of both fields. Studies should be initiated by each group, and evaluations should be accomplished by multidisciplinary teams consisting of both social scientists and designers. The leadership of such teams can vary according to whether the research is primarily addressed to social science concerns or to design issues.

Yet, while the skills and talents of both social scientists and designers are essential to evaluative studies, a number of authors have noted the difficulty of relating the inherently different approaches they employ to common objectives (Altman, 1973; Lang and Moleski, 1973; Ostrander, 1974; Rapoport, 1969). For example, Altman (1973) suggested four general differences in the scientists' and designers' approach. He suggests that the typical environmental designer usually studies a particular unit or place such as a home or neighborhood, while the typical behavioral scientist studies a particular phenomenon or social process such as privacy, territoriality, or crowding. A second difference he notes is that the orientation of the designer is toward the final product and that the orientation of the scientist is toward information-gathering processes. A third difference is that of the synthesizing nature of the designers' activities and the analyzing nature of many of the scientists' activities. While one is putting things together, the other frequently pursues a study of the effects of individual variables on specific behaviors. The fourth difference which Altman suggests is that of the "doing and implementing" nature of the designers' responsibilities and the "knowing and understanding" thrust of the scientists' interests.

Ostrander (1974) has suggested that the differences in the modes of communication also may be an important distinction between the designer and the scientist. He suggests that the designers' reliance on visual modes of cognizing and communicating and the behavioral scientist's reliance on semantic modes represent two distinct professional cultures and that these differences may create conditions of stress when communication is attempted.

Although these issues have not been solved, there are a growing number of successful collaborations between designers and social scientists—at least we can see the tunnel, if not the light at the end of it. Although there have been different patterns used in these successes, some common elements seem to characterize many of them. Social scientists have learned to be somewhat more visual; designers have learned to be more expressive in writing. Similarly, common goals have been expressed "bilingually": verbally and visually. Moreover, evaluations which share a common theoretical/conceptual base (e.g., a desire to understand privacy) seem more successful. Most of all, these studies have shown patience and the ability to compromise. In such collaborative studies, compromise often turns into synergy.

What is to be evaluated? Activities and responsibilities of designers and planners encompass a broad array of environmental settings. These

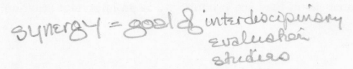
synergy = goal of interdisciplinary evaluation studies

range from small scale to regional scale; they include public and private clients, and they serve individuals and populations. Where does one start in sampling this array to initiate evaluative studies?

Our discussion of the need for evaluation suggested some general criteria for establishing priorities for evaluation: specifically choosing settings of applied and/or theoretical significance. These criteria imply that several kinds of settings are of high priority for evaluation:

1. Settings where the accountability of the program, designer, or client is an important issue, such as:
    a. Settings which are constructed or otherwise subsidized totally or in part with public funds;
    b. Settings such as factories, work places, prisons, hospitals, and psychiatric institutions, where the users normally have little input into the design process or control over their lives;
    c. Settings, either publicly or privately financed, which are intended for the use of the public;
2. Settings which affect many people and are likely to be frequently replicated, such as mass housing or commercial offices;
3. Settings which vary on important theoretical dimensions, such as settings which differ on their "opportunity to control personal experience" (Knight *et al.*, 1977), on their "thwarting of personal needs" (Stokols, 1976) or on their basic aspects of quality of life (Craik and Zube, 1976).

The Evaluation Process

A major theme of this book is that the design process must be a cumulative process in which designers learn from their successes and mistakes and subsequently improve their designs. Indeed, it is useful to think of the design process and the evaluation process not as separate steps but rather as a single unit: the design–evaluation–design cycle. This cycle is illustrated in Figure 2.

What is the focal problem? Every evaluation contains some relationships of special concern; these form the focal problem. In some cases the information user or the sponsor of the evaluation may define the focal problem. For example, a government agency may be interested in evaluating the impact of housing regulations (design activity) on the satisfaction of public housing residents (users). A design firm may be primarily interested in understanding how certain materials or spatial arrangements (setting characteristics) affect office workers (users).

There may often be *conflicting* pressures to define the focal problem. Management may see productivity as an important criterion, whereas

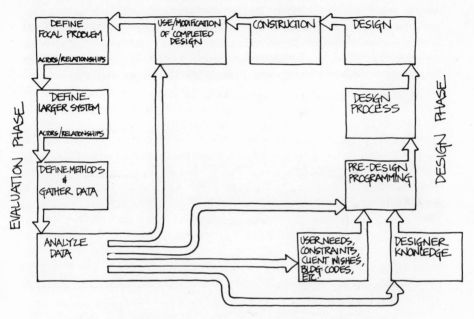

FIGURE 2. The design-evaluation-design cycle.

the union may see comfort or satisfaction as more important. A federal agency may see the impacts of public policy on the form of the setting as a prime issue, whereas the evaluator may feel that theoretical links between the setting and user behavior are more important.

Also, serendipitous events may suggest a focal problem. A large firm might move into a new office, a hospital might move or renovate, a new building may be built. Such events provide opportunities for evaluations which can be of considerable applied or theoretical importance.

In addition to the external influences, the functioning of the setting itself helps to define the elements that must be considered in the focal problem if the evaluation is to be useful. For example, in an evaluation the relationship of most concern to the evaluation sponsors might be between a new office arrangement and secretarial efficiency, yet the functioning of a secretary is dependent on his or her supervisor; hence office managers must be considered as part of the focal problem. In a hypothetical example previously discussed, the relationship of interest was between public housing regulation and user satisfaction, yet research has shown that upkeep has a significant impact of satisfaction. So maintenance must be considered as a part of the focal problem.

The focal problem can be simply conceptualized as involving two components: the *elements* of interest and their *interrelationships*. The elements are the user groups, physical features, or social influences that make up the focal system; their interrelationships are, of course, the ways we expect them to affect each other. In the example of office design mentioned above, the new office design, office managers, and secretaries are elements in the focal problem. The ways that they are expected to influence each other include such interrelationships as an open-plan design increasing social interaction by clustering people, better surroundings increasing morale, and so on.

The ELEMR evaluation in Chapter III of renovations at an institution for developmentally disabled people exemplifies explicit choice of the focal problem (Knight *et al.*, 1977). The project was funded by the Developmental Disabilities Office, an HEW agency with responsibilities for establishing environmental design and programmatic guidelines for institutions for the developmentally disabled. Even though the evaluative factors of prime interest were therapeutic changes in the behavior of the residents, observation in the setting quickly revealed that the direct-care staff were all-powerful in determining how the residents would use the renovations. Clearly, the staff had to be considered as part of the focal problem.

The elements were the renovated environment, the staff, and the residents. But what were their relationships? Observation in the setting, informal interviews, and background reading revealed several ways in which they interacted (Figure 3). The renovations provided spaces, doors, light switches, and so on. These amenities provided the opportunity to model and teach appropriate behavior for residents. In the same sense the physical environment offered the same opportunities to residents: for the first time residents could potentially make a room darker or brighter, noisier or quieter. Yet the staff determined whether these opportunities were realized. It was the staff who allowed (or did not allow) the residents to interact with their environment.

The study of a campus open space by Cohen *et al.* (1976)(see Chapter IV) was less complex. Since the evaluators were trying to provide information about student perceptions and behaviors for an upcoming design of the space, the elements in this evaluation were simply the students and the setting itself. The relationships of interest were between the students' attitudes and perceptions of the space and the form of the space itself, and between student behavior and the form of the space.

In summary, there are several factors which help define the elements and interrelations which form the focal problem:

1. Needs and values of the users of the design.

2. Needs and values of the sponsor of the evaluation (government agency, design firm, design school).
3. Background, needs, and values of the intended users of the information (designer, social scientist, policymaker).
4. Background, interests, and goals of the evaluator.
5. Functioning of the setting itself (who interacts with whom, who has power over whom, etc.).
6. Opportunities provided by a new or altered design (a new park, a new building).

What is the larger system? The focal problem is the primary concern of the evaluation, yet each focal problem is affected by many other direct and indirect influences. The focal problem may inject only two factors into the appraisal—perhaps the setting and the users—but the relationship between these factors may be affected by the social–historical context, the proximate environmental context, and/or the design process. In other words, when we specify the larger system, we are elaborating influences on the focal problem which are not the critical issues in the evaluation, yet which are important in understanding them. For example, consideration of the larger system might involve asking questions such as: Would a poor neighborhood decrease user satisfaction for an

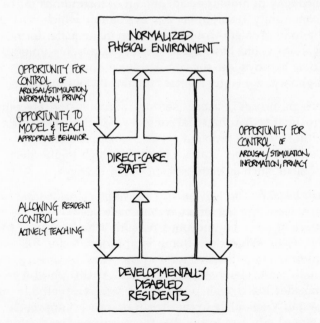

FIGURE 3. The focal problem for the ELEMR Project. The size of arrows reflects approximate strength of influence.

apartment complex or a nice neighborhood increase it? Has an economic change meant that there are better (or worse) teachers staffing a new school and hence a new design is used differently? Has an administrative policy change made employees more (or less) satisfied with their jobs, regardless of the effects of the design to be evaluated?

Like the focal problem, the larger system may be thought of as having two components: *elements* and *interrelationships*. In the previously mentioned evaluation of a new office design, the focal problem was the ways in which the design affected secretarial efficiency. This focal problem would be influenced by such larger-system elements as the policies of the upper management, the activities of the union, and by the local and national economy (which might make secretarial jobs easier or harder to get). All of these influences could tend to alter the impact that the new office design has on efficiency. While it is impossible to consider all possible factors that might constitute the larger system, it *is* important to list the most influential impacts on the focal problem.

In the ELEMR Project, the residents, environment, and staff formed the focal problem, yet this triad was directly affected by the administration, the professional staff, the parents' association, and so on. This larger system is illustrated in Figure 4. In addition, more subtle factors needed to be considered in the evaluation, such as the prevailing treatment philosophy of moving residents out of institutions into a range of smaller community residences. An evaluation which had examined simply the focal problem would have been inadequate, for no effects of the renovations could have been considered positive unless they prepared people to move into the community.

In summary, the larger system contains factors such as:

1. Design process, users, setting, social–historical context, and proximate environmental context when these factors do not constitute the focal problem yet affect that problem.
2. Issues at a larger scale of analysis which might affect the focal problem, such as management policy changes.

How do you define the appropriate methods for an evaluation study? After the focal problem and the larger system have been defined, some of the most important research questions remain: What sampling procedures should be used? What time frame is appropriate for the evaluation? Which information-gathering techniques (such as direct observation, questionnaire, interview) should be used? At this point it is useful to briefly consider the overall purposes of research methods and to see how these purposes relate to the structure–process approach.

It is almost a truism to say that the purpose of research methods is

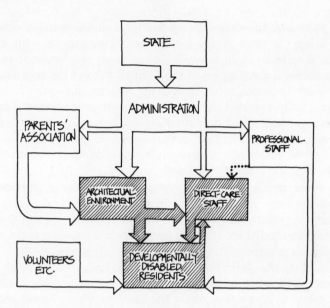

FIGURE 4. The focal problem (shaded) in relation to the larger system for the ELEMR Project. The size of arrows reflects approximate strength of influence.

to provide "useful" information. Yet what makes information useful? It is apparent that it must be appropriate. It must be concise, clearly stated, and straightforward; it must answer the critical questions for the user of the information.

It is perhaps less apparent that the information must be of "good quality": it should show some consensus between observers; it should describe the conditions, time, setting, etc., well enough to understand where generalization is possible; it should be sensitive to small changes in the most important issues of the evaluation and to unexpected changes in other issues; it should correctly attribute the influences of the observations. Indeed, these issues are the central concern of all social research, as well as of environmental design research.

The evaluator influences the quality of his or her evaluation by making two sorts of decisions: (1) questions of study design, such as which settings should be studied, who should be sampled, and what time intervals should be used; and (2) questions of specific techniques, including choosing among qualitative or quantitative methods, interviews or questionnaires (or both), direct-observation techniques, and so on. Although these questions appear complex, carefully defining the focal problem and the larger system helps provide answers which increase the usefulness of the resulting information.

For example, the definition of the focal problem helps select techniques. Since the focal problem consists of two aspects—the elements and their interrelationships—techniques must be chosen that directly measure these. For example, in the ELEMR Project the relationships of interest were between renovated living spaces, the mostly nonverbal developmentally disabled residents, and the direct-care staff. These relationships were tapped by documenting the nature of the environments, directly observing residents and staff–resident interactions, interviewing staff about residents, and recording the location of both staff and residents over the course of the day.

Since the needs of various groups, such as users, sponsors, and information users, are important in defining the focal problem, it is important to *explicitly* involve them in the definition of the focal problem. This helps to insure that the necessary information is collected for every relevant group./

Of course, the definition of the larger system also helps suggest information-gathering techniques. The larger system was defined because it could in some way affect the focal problem, hence techniques must be chosen that measure the interrelationships of the larger system with the focal problem. For example, when Zube *et al*. (1976) evaluated twelve visitor interpretive centers in National Parks, their focal problem was the effectiveness of the center designs and of the interpretive programs for visitors (see Chapter III). The evaluators primarily relied on interviews and direct observations of the visitors to address the focal problem. The evaluators were also concerned with such larger issues as increased park use due to changed attitudes about the outdoors, increased crime, and structural problems caused by extreme weather. Judicious use of other information-gathering techniques, such as interviews with staff members and analysis of records, helped to understand these issues. Hence, concern with the larger system often leads to a greater depth and variety of data-gathering methods.

Defining the larger problem and gathering relevant data can also increase the quality of information. The careful definition of a setting provides a basis for comparison with other settings and helps to suggest which generalizations may be made. It suggests which relationships must be monitored for unexpected changes and helps in labeling the effects measured in the evaluation. If many relationships are being monitored, we can be more confident that we are correctly labeling the effect. For example, in the Brookes (1972) study of a landscaped office (Chapter II), the evaluator probably correctly labeled the increased preference he measured as due to the design of the new office, yet his results would have been more convincing if he had documented pay raises, management changes, and other issues that affect satisfaction.

It is clear that the structure–process approach requires choosing *several* information-gathering techniques, including different qualitative and quantitative methods. Methods sensitive to small changes must be used to monitor the focal problem while method sensitive to unexpected changes must be used to monitor both the focal problem and the larger system.

In short, we have suggested that the role of methods in an evaluation is to gather useful information by properly designing the study and by choosing appropriate techniques; the definition of the focal problem and of the larger system aids the evaluator in his or her effort. A likely conclusion of this process is that the evaluator will choose a mix of qualitative and quantitative methods.

How do you analyze the data? Data analysis can be a complex, sophisticated process which requires years of training and necessitates access to high-speed computers. In fact, the analyses for several of the cases discussed in Chapters II, III, and IV are of this nature. Nonetheless, important and valuable insights can be learned by very simple analytic methods. The principle underlying *all* analytic methods is *understanding*. No matter how simple or complex, data analysis should aid in understanding the structure and the relationships present in the data. If analysis is not understood, the danger of making errors is dramatically increased.

A general rule for analyzing data is to progress from simple to complex. A first step is to lay out the important issues in the data and list the major points raised in interviews. Then when it is appropriate, plot percentages and draw graphs.

For example, in the study of Bryant Park in midtown Manhattan, the evaluators used bar graphs to compare the ratings of the park given by people according to age, sex, and frequency of use. In the ELEMR study of an institution for the developmentally disabled, the evaluators gained insight into the changes due to renovations by plotting graphs of behavior of the residents over the several observations before and after renovations.

It is of central importance to work with the data until they become familiar and understood. This is especially important with the multimethod approaches which we are proposing, where the workings of the focal problem and the larger system can only be understood by combining the results of different methods. For example, in the study of visitor interpretive centers, Zube *et al.* (1976) used interviews, direct observations, and other techniques to evaluate the settings. By weaving together the results of these techniques, the evaluators could produce a whole cloth that described the focal problem and the larger system.

Once the data are understood at their simplest level, evaluators can analyze on a more complex level. However, evaluators seldom attempt more sophisticated statistical analyses, which is, in our view, a shortcoming. Since humans are naturally variable in their behavior, merely listing percentages does not consider the *variability* of people's actions that more sophisticated techniques take into account. Individuals affected by mood, weather, or health may react differently to the same situation. Sophisticated methods allow researchers to understand whether their findings are likely to be real or are due to natural variability.

For example, Brookes found in his study of an open-plan office that the users rated their conventional office 1.4 on a 5-point scale of utility (1 = least, 5 = most), while they rated their open-plan office 1.8. While this difference appears small, Brookes noted that the ratings were very consistent. Since almost all raters marked their conventional office between 1.4 and 1.6 and their open-plan office between 1.7 and 1.9, Brookes was confident that this was a real difference. By contrast, had all of the ratings varied from 1 to 4, with the open-plan office simply receiving a few more 4s, it would have been plausible to assume that, by chance, the open-plan group had simply included a few more enthusiastic people. With another group of people he might have received a few less enthusiastic responses which would have changed the ratings. Low variability in estimates provides a sounder basis for interpreting differences.

How can we feed evaluation information into the design cycle? In this chapter, four major strategies have been suggested to increase the flow of evaluation information into the design cycle. First, it was suggested that a primary responsibility for this rests with the design schools by teaching the importance of design evaluation, how to use evaluation information in the design process, and by actually using information derived from real settings. If students gain the "evaluation habit" and see that it improves their designs, they are more likely to use it in their professional practice.

Second, it was suggested that a more comprehensive data bank of evaluation information could be established, perhaps by collaboration between design schools and the relevant professional organizations. These data could be organized by setting, users, context, etc., and would be accessible to both professionals and students.

Third, design firms who evaluate their own work or who collaborate with social scientists in evaluating their own designs may be more likely to produce information that can be fed back directly into the design cycle. In Chapter II, the REDE group evaluated their own design of a college dormitory renovation. Also in Chapter II, Reizenstein *et al.*

formed a multidisciplinary team of designers, planners, and social scientists who oversaw the entire cycle from programming to design evaluation.

Fourth, the information must be useful if it is to be put into the design process. Usefulness has two senses: "Is the information appropriate?" and "Is the information of 'good quality'?" The first question may be addressed by careful definition of the focus problem, perhaps with participation by the eventual user of the information. The second question can be aided by careful definition, by cognizance of the larger system, and by the appropriate choice of methods.

Finally, information must be made applicable to all the participants in the design process: designers; policy planners at the local, regional, and national levels; builders; owners and developers, and bankers. This suggests that evaluation must address a broad range of needs, such as user satisfaction, life-safety issues, maintenance, and management concerns.

SUMMARY AND CONCLUSIONS

We have suggested that environmental design evaluation consists of a diverse set of studies and practices with a common objective: assessing the effectiveness of a designed environment for users. Although it is possible to usefully spend much time and money on an evaluation, a short and inexpensive evaluation may be quite useful if it is carefully thought out and well designed.

A two-faceted structure–process approach was presented: (1) a five-part conceptual *structure* and (2) a multistep *process* for actually accomplishing the evaluation.

The conceptual structure provides a way to organize the information required for an evaluation. It consists of (1) the setting, (2) the users, (3) the proximate environmental context, (4) the design process, and (5) the social–historical context. The evaluation process provides an overview to define the larger system, choose methods, analyze results, define the focal problem, and feed back the evaluation information into the design cycle.

We have presented a large number of evaluative factors and have suggested several steps in the evaluation process, each of which involves several considerations. No single evaluation will empirically address every evaluative factor or answer every concern in the evaluation process. It is our contention, however, that even a modest evaluation can be useful if the evaluators consciously address the questions raised in this chapter. The following 3 chapters present 14 studies which illustrate a broad range of approaches, settings, and measures. These detailed case studies amplify the principles discussed in this chapter and provide an overview of current work in environmental design evaluation.

II

Interior Spaces

Interior spaces surely affect people more strongly, certainly more directly, than outdoor spaces or buildings-as-architectural-entities. Yet for a number of reasons interior spaces have received relatively scant attention from researchers in the field of design evaluation. One of the reasons is the fact that interior design has only emerged as a truly professional field in recent years; however, the field in general (and particularly nonresidential interior design) has now taken its place alongside other serious design professions. Also, many people exhibit a kind of possessive attitude toward their interior spaces which is an extension of the concept that "my home is my castle." It is difficult enough to evaluate such spaces as offices, shops, and other business interiors without violating a kind of privacy or private-space territoriality imparted to those spaces by their occupants. It is almost impossible to conduct a truly objective study of personal spaces such as homes. Yet it seems obvious that one's most personal space—the home—is precisely the space that affects behavior, well-being, and general satisfaction (or dissatisfaction) more strongly than any other part of the built environment.

The state of the art in the field of interior design evaluation is somewhat limited. It is safe, however, to predict that the next few years will see a tremendous increase of meaningful evaluation studies of all kinds of interior spaces. Undoubtedly it will also become possible to research and study a variety of home settings, and it is reasonable to expect that a great deal will be learned about the real meanings of space, form, color, texture, etc., in relation to their influence on behavior.

The studies included in this chapter consist of a college dormitory, offices in a business firm, offices in a hospital, a new Fine Arts Center, and several New York City subway stations. Perhaps the latter is not an interior space in the conventional sense of the word, but it certainly is a totally manmade environment affecting thousands of people each day.

One of the most significant aspects of interior design evaluation is shown in several of the studies, but in particular in the one of an office

space by Malcolm J. Brookes. It is the fact that in an interior, the evaluation process can easily become a process of change and modification. It
would be costly and difficult to undertake major structural alterations of
buildings in response to research studies; one can hardly conceive of
major modifications in communities, outdoor spaces, and landscapes.
The nature of renovation and remodeling, however, is not costly, and,
of course, that is what much of interior design is all about. Hence the
kind of example given in the Brookes study is an exemplary direction
that might easily be adapted by other designers/researchers.

The settings discussed in this section vary considerably in scale. The
Social Services Office in the Cambridge Hospital is moderate in scale
and scope, while the subway stations consist of very large spaces which
in turn are just segments of a vast interrelated network of spaces. The
settings also vary considerably in their purposes, and the study designs
had a number of different objectives. Although the basic concern of this
book is postconstruction evaluation, the example of predesign evaluation
(or preredesign evaluation) of the subway stations points out the essential need for programming based upon thorough research and evaluation studies. Likewise, the study dealing with the Fine Arts Center at the
University of Massachusetts clearly puts the blame for shortcomings (if
indeed there need be a "villain" or a focus for blame) on the lack of
proper programming. The evaluation of Butterfield Hall Dormitory
undertaken by REDE was an ongoing observation which led to a certain
amount of design decisions and modifications in the process of the actual renovation. But more importantly, REDE has already applied the
information gathered from the Butterfield study to the design of dormitories at Brown University, the University of Wisconsin at Milwaukee,
and several other design projects dealing with dormitories.

The predesign programming aspect of design, as well as the evaluation of completed jobs as a feedback mechanism for future design, are
two extremely important facets for any design. Information learned from
evaluation studies in the cases discussed here was factual information
gathered and needed for several aspects of design decision-making.
These were not theoretical studies undertaken in the pursuit of abstract
knowledge. Perhaps the only project that did not utilize the findings and
recommendations of the research was the subway study, although it is
likely that at some future date the information may be used and applied.
The record of subway building in recent years, however, especially in
New York City, shows it to be a very slow process; hence, it would be
unusual indeed if the results of this study had been used immediately.

It follows, then, that evaluation studies are real tools for the designer. They are of such obvious utility that it seems very clear that they

bring immediate benefits to the designer and client alike. Not every design commission is very large, with a sizable fee for the designer; hence the example of the Social Services Office of the Cambridge Hospital was selected for inclusion in this section. It was a very low budget job, dealing with only about 2000 square feet and minimal furnishings. It exemplifies that evaluation studies can be carried out by every designer in any size job without large expenditures and large research organization.

The question of objectivity might be raised in connection with this study and the one carried out by REDE since both studies were done by the design firm itself. In both cases, however, behavioral scientists were part of the team, allowing a separation of specific tasks and aiding the studies to be nearly as unbiased as similar studies done by outside evaluators. But more importantly, the fact that design firms have started to carry out evaluations and use the results for modifications and future design projects is precisely the proof for the validity of design evaluation as a meaningful tool.

Interior design evaluations need to focus accurately on the problem to be evaluated and even on a specific aspect of the problem. The need for the definition of focal problems was pointed out in Chapter I. Because these studies deal with the variables of specific individuals and their interaction with their environment, focal problems must be clearly spelled out before an evaluation is undertaken. The dormitory study in this chapter set out to evaluate student satisfaction. It could have addressed the problem of financial returns, occupancy rate, maintenance, vandalism, crime, or the role of a dormitory as an extension of the learning process. The study dealing with an office landscape was designed to find out employee reaction to such a system. The focus might have been efficiency from the management's point of view (although this includes employee attitude), energy consumption in terms of heating, ventilation, air conditioning, and lighting, or simply productivity.

The Fine Arts Center study dealt with symbolism, aesthetics and perception, rather than focusing on the quality of the building's teaching spaces, acoustics in the concert hall, or function of the art gallery as an exhibition space. The Cambridge Hospital office study addressed the satisfaction of the staff members and the function of the offices. It did not focus on concerns that might have been expressed by the hospital administration. And finally the subway study is primarily a predesign evaluation focusing on pedestrian movement in several stations rather than addressing the myriad problems that are inherent in a complex mass transportation system.

The need to consider the ever-present social–historical context was pointed out in the first chapter. We are usually aware of such a context without taking conscious recognition. Until a few years ago, office design, for instance, took for granted highly structured hierarchial approaches. The size, location, materials, and furnishings of an office were as clear an indication of the occupant's status as the title on the door. Although this is still the case in many corporations, the office landscape system discussed in the Brookes study views the actors in an office environment in a totally different social context. This office landscaping would not have been an acceptable design approach just two decades ago.

The rules governing dormitory living have changed drastically in the past few years. Hence the social–historical context of dormitory design and evaluation in the mid-70's is based on underlying mores and assumptions which would have shocked college administrators in 1950.

The literature of environmental psychology and design evaluation is full of examples proving that the designed environment alone cannot determine human behavior. It is clear, however, that the proximate environment is indeed a strong contributor to the direction of human behavior; therefore, evaluations of interior spaces that recognize the many interactions of social variables, but focus clearly on specific aspects, can provide useful and significant data for designers. Perhaps evaluations of interior spaces, more than any other aspect of the larger field of man–environment relations, represent the most clearly tangible results which can provide the answers sought by design professionals and others commissioning environmental design evaluations.

Changes in Employee Attitudes and Work Practices in an Office Landscape

Study by: Malcolm J. Brookes
Methods used: Questionnaires (semantic scales and open-ended questions) and interviews
Type of project: Study of the effects of changes in office environment from a conventional office to an office-landscaped one
Information source: *Human Factors* 1972, Malcolm J. Brookes and Archie Kaplan

INTRODUCTION

The design of offices is a significant element in United States business. A major alternative to conventional rectilinear planning was introduced in the early 1960s. The innovative concept was introduced by West Germans under the name of "Bürolandschaft." Offices are planned around the organizational processes within the business, and people who work together are physically grouped together. Fixed partitions, cubicles, and rows of desks are eliminated in favor of one large open space reflecting the patterns of the work groups. Spaces are broken up with screens and plants, and one of the key concepts of the system is the fact that all staff members participate in the open plan, not just clerical workers and a few supervisors. The psychological need for privacy is accomplished with dividers and placement of furniture. Since the office landscaping does not use fixed walls, it is possible for various departments to change their work areas, reflecting changing needs of business.

Claims have been made that the open system provides greater efficiency of staff, more satisfied employees, more economical utilization of space, and more economical upkeep over a period of years. Controversy between advocates of conventional office planning and office landscape planners is still prevalent today.

The study undertaken by Malcolm J. Brookes was one of the earliest thoroughly structured and objectively designed research projects in the field. Its focal problem was the study of employee attitudes toward their work environment through a carefully planned "before and after" evaluation.

A major U.S. retail firm commissioned the study prior to the design of new headquarters offices for several thousand staff members. The management wanted to know whether a landscape style of office would be preferable for their operations. It was decided to actually build a pilot project. Three departments, staffed by 120 people (purchasing, customer service, and building services management), occupied roughly 30,000 square feet in a conventional rectilinear office space. Approximately half of the participants were female. Questionnaires were administered at the outset of the project. The group was then moved out to temporary offices until their old space was redesigned, taking into consideration some of the findings of the first questionnaire. The new design used office landscaping. Nine months after the group was moved back into the redesigned offices, a second set of questionnaires was administered (85% of the original staff members remained).

STUDY DESIGN AND METHODS

The principal research instrument was a questionnaire consisting of 45 semantic scales consisting of connotative (e.g., "hard, hostile") and denotative items (e.g., "noisy, light"). A modified Stapel scale was used, rather than the more familiar bipolar adjective checklist. In the Stapel scale, the responder is asked to rate a single adjective on a 5-point scale reflecting its association to the overall concept (from 1, low association, to 5, high association). For example, responders were asked to rate the degree of association between the adjective "noisy" and their existing office. Average ratings are presented in Figures 5–8. A line extending to the left indicates low association, to the right, high association. Length differences between two bars adjacent to an adjective reflects differences in ratings.

The questionnaires were presented at two times: prior to formal announcement of the renovations (but after rumors had circulated) (Figure 5), and 9 months after the office workers returned to their renovated offices (Figure 8). At each administration of the questionnaire the responders were asked to check the scales in a manner which reflected their reactions to (1) their existing workspace, (2) their ideal workspace, (3) their co-workers' ideal workspace. Comparing the ratings of existing unrenovated workspace with the ideal workspace helped highlight exist-

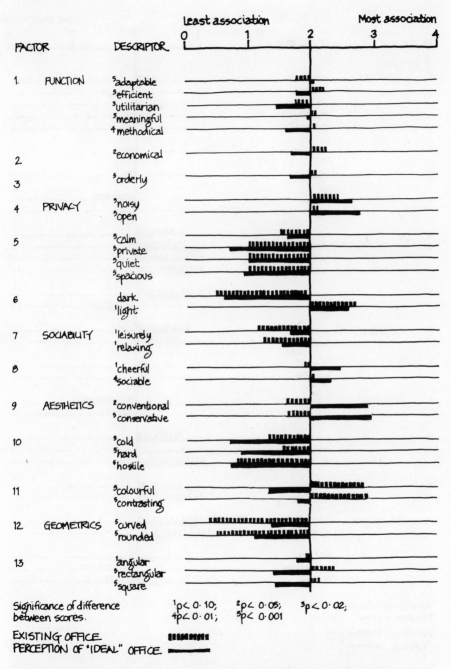

FIGURE 5. Responses to the conventional office and to perceptions of an "ideal" office on semantic differential scales. A word on the left (e.g., "function") corresponds to a name for a factor or cluster of scales.

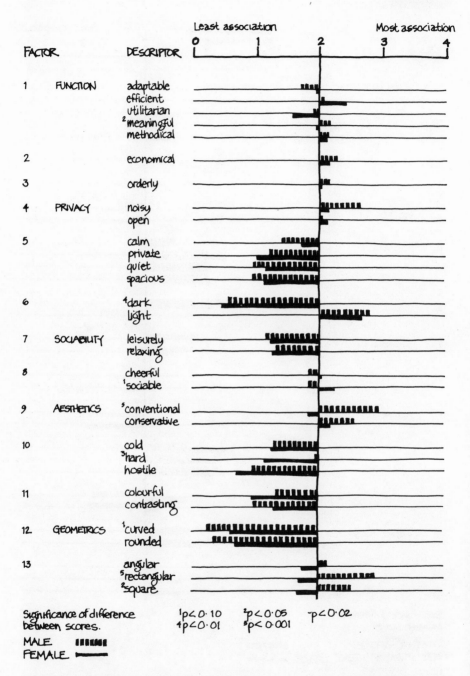

FIGURE 6. Responses to the conventional office on semantic differential scales compared according to sex of respondent.

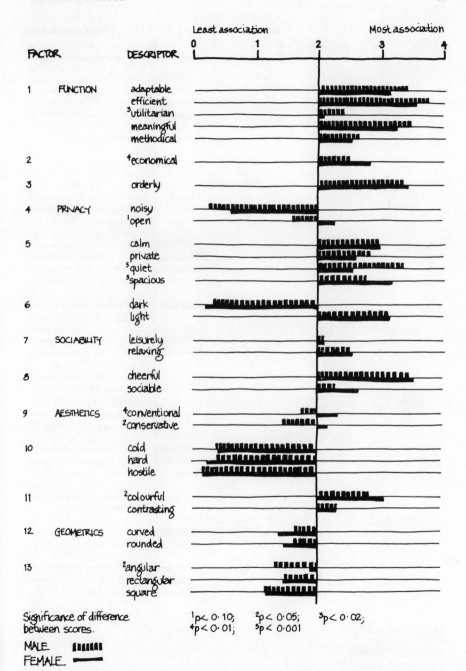

FIGURE 7. Conceptions of an ideal office on semantic differential scales compared according to sex of respondent.

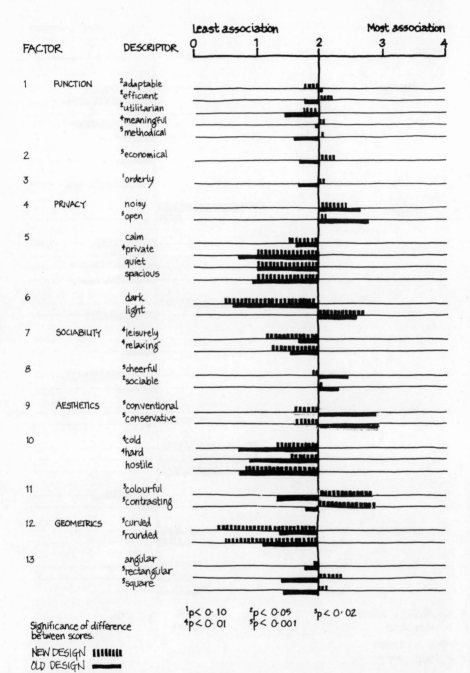

FIGURE 8. Perception of new office compared to old office on semantic differential scales.

ing strengths and weaknesses and provided information for the design of renovations.

The 45 adjectives were an unwieldy number to analyze, so factor analysis was used as a data-reduction technique. Factor loadings were produced after rotating the factors to a varimax rotation. After examining the factors produced from several analyses, the authors concluded that the 45 adjectives could be described by 13 underlying factors (left-hand numbers on Figures 5–8) which remained relatively stable (60% of the variance was accounted for). These factors are indicated in Figures 5–8 and include dimensions such as function, privacy, sociability, aesthetics, and geometrics.

It was a prime concern of the evaluation to determine whether the ratings of the unrenovated and renovated environments actually differed, or whether differences were due to chance. To do so, chi-square analyses were performed on contingency tables dividing the data based on various characteristics such as sex, department, instructional set (existing, before, after) and time (before, after). As Figures 5 and 8 illustrate, most differences were "significant" (i.e., we can be quite confident that they were real and not due to chance). For example, there were many real differences in ratings between the unrenovated and renovated offices, the unrenovated and ideal offices, and the renovated and ideal offices.

In addition, each of the participants was individually interviewed before and after moving into the new office. Finally, each participant was asked to list the three items that they most disliked about the unrenovated and renovated offices.

Setting and Proximate Environmental Context

The existing space was one floor in a typical office building. It was laid out in a rectilinear open plan, with the three departments separated by corridors formed by fixed partitions. Some of the executives had private offices and some supervisors had cubicles partitioned off by head-high dividers. Clerical workers and secretaries were arranged in rows of desks in their respective departments. The furniture was conventional grey and green, and the floor was tile. The new offices had dark-colored, wall-to-wall carpeting, new furniture, plants, bright colors, and movable screens.

Space planning—the layout, organization, and design of offices—has taken on tremendous importance in the last 20 years. In 1970, nearly half of the country's civilian labor force, totaling 79 million persons, were white collar workers. Hence, the claims made by observers of European installations of office landscape systems (predating the introduction of such systems in the United States by several years) attracted a great deal of attention among American business executives. Some of the claims in favor of office landscaping stated that it resulted in:

40–50% reduction in space requirements
20% decrease in maintenance costs
95% reduction in set-up and renovation times
10–20% increase in staff productivity
Improved staff morale and decreased absenteeism

None of these claims has been substantiated by advocates of office landscaping, nor have opponents been able to disprove the claims. Many variables, such as staff morale caused by general job satisfaction and income, have not been examined in this particular study and are always difficult to determine. Opponents of office landscaping point out the higher cost of equipment and furniture and claim that a conventional installation is less costly to build and maintain.

Prior to undertaking the study reported here, the researchers conducted interviews with representatives of ten companies who had installed office landscapes, but the results were inconclusive. It was in this context of conflict between conceptual assumptions and unsubstantiated claims on both sides of the controversy that the research project was undertaken.

EVALUATION

The answers as to whether the new offices function better than the old ones can be extrapolated from Figures 6 and 8. In terms of sociability, the responses were positive; but on the negative side, the new offices were found to be more noisy and bustly and there was a marked loss of privacy. Brookes raises the question whether privacy and security require visually solid barriers—that it is not enough to "know" that you cannot be overheard. A comparison between the old office and the resulting design (Figure 8) shows that the office was improved on some factors, but that some negative things still appear: "cool, hard and hostile." Lighting was perceived as unchanged, although in reality it had increased in the new offices. This wrong perception might have been due to the dark color of the carpeting compared to the former tile floor. In Brookes' words:

> In personal space control, noise, privacy, etc., the landscape design falls short of its goals; and it is disappointingly inefficient. But whether this is due to management policies and employee attitudes—both notoriously conserva-

tive in this instance—or the architectural design, is not made clear by this study.

One of the most important elements in the evaluation was the part dealing with open-ended questions asking respondents to list three major things they like and three things they dislike about the old (existing) office, and subsequently the identical questions asked about the new offices. Items on the lists are ranked according to frequency of mention. The two listings are reproduced here and make for some very interesting comparisons. (See Tables 2 and 3.)

DISCUSSION

As quoted in Brookes' report, Manning (1965) wrote:

> At present, design decisions affecting the social environment of office buildings are made almost entirely on the basis of expectation or personal prejudice, rather than knowledge.

That statement is still very true today, not only relating to social environments of office buildings, but to most aspects of interior design. One

TABLE 2
Old Conventional Office: Likes and Dislikes[a]

Dislikes	Likes
Crowding of desks, 47	Good lighting, 37
Noise of conversations, 43	Good adjacencies and layout, 34
Poor adjacencies and layout, 37	Good furniture and equipment, 27
Poor HVAC, 31	Adequate privacy, 27
Poor lighting and glare, 24	Sufficient room, 16
Drab decor, 21	Location in building, 11
Inadequate workspace/storage/ shelving, 20	Pleasant co-workers, 11
Inadequate filing spaces, 20	Cleanliness of workspace, 7
Lack of privacy, 18	Decor, 7
Lack of provisions for visitors, 13	Close to canteen, 7 (see above, location in bldg.)
No windows, 8	Adequate HVAC, 6
Lack of machines and business equipment, 8	Efficient atmosphere, 4
Need for coat racks, 5	Quiet, 2
Lack of flexibility in arrangement of space, 3	Other people's offices, 2
Lack of room for expansion, 1	Impression made on visitors, 2
	Good pin-up space, 1

[a]The frequency of responses to an open-ended question asking subjects to list the three major things they like and things they dislike about the old conventional office. Items ranked according to frequency of mention.

TABLE 3
New Landscaped Office: Likes and Dislikes[a]

Dislikes	Likes
Noise of conversation, 47	Colorful design, 34
Lack of privacy, 42	Comfortable chairs/furniture/
Crowding of desks, 43	workplace, 34
Flat-top desk too small, 26	Attractive decor, 26
Lack of drawer space for personal	Indoor plants, 18
storage, 24	Good lighting, 18
Adjacencies and layout, 24	Cheerful atmosphere, 13
Poor HVAC, 17	Carpet on floor, 13
Lack of shelf space, 16	Lack of noise, 12
Poor lighting, 8	Adequate adjacencies and layout, 11
Too many plants and insects, 7	Modern appearance, 10
Cheapness of furniture construction, 7	Adaptability of furniture, 7
Lack of filing space, 7	Adequate privacy, 6
Drab decor, 6	Small conference areas/provisions
In traffic flow, 5	for visitors, 5
Uncomfortable chairs, 4	Ease of keeping workplace clean
Lack of electrical outlet, 3	and tidy, 4
Getting lost, 2	Better filing space, 4
No windows, 2	More personal contact, 4
No provision for visitors, 2	Good impression on visitors, 3
Poor phone location at desk, 2	Out of traffic flow, 2
Poor expression of status, 1	Easy to supervise secretary, 1
No buzzer on phone, 1	

[a]Nine months after moving to the new landscaped office the subjects reported the following major "likes" and "dislikes." Items are ranked according to frequency of mention.

might speculate on whether interior environments can ever (or should ever) be designed completely objectively. Those who consider interior design (and architecture) an art would probably insist that the designer's intuition should never be replaced by "scientific" solutions.

Malcolm J. Brookes' study is, of course, an excellent example of a very thorough and scientific study, yet carried out with sensitivity to design and with real professionalism. It is significant as one of the very few objective research projects in the field of space planning and interior design. It had to be done particularly in relation to "Bürolandschaft" since its proponents have for years assumed a scientific basis for their claims, without any quantifiable data.

The object of the study was to research productivity, group cohesion and interaction, morale, and attitudes to work and environment in the old offices, and then to examine what changes had taken place in the new offices. Unfortunately, it was not possible to include some kind of personality metrics and demographic data. The study was commissioned by the clients; hence, the researcher had to accept limitations imposed by them.

The report mentions personal "in-depth" interviews in connection with the open-ended questionnaire, asking for things liked and disliked. Yet, the results of these interviews were not reported. It would have been interesting to read the author's comments and findings about these interviews and glimpse somewhat more information about the nature of the questions and responses.

The study has one truly exemplary aspect: It dealt with research and *modification* or *change* based upon the research. Evaluation studies only rarely can implement the findings of their research, and indeed it is not always the goal and function of such studies. However, change and modification of the environment to implement findings or to test out a hypothesis is possible only in interior design. It would be too costly to make major architectural changes in an existing building and likewise in large-scale planning projects. Hence this project came closer to what the field of environmental behavior or environmental psychology is all about. It is a study which points the way to a truly constructive direction of research and actual implementation. Too many studies in the field, and too many evaluations, have been done and published, only to gather dust upon the shelves of other researchers. It is, in summary, an excellent project, one that other designers might be able to take as a pattern for the positive potential of design evaluation.

Butterfield Hall: Evaluation of a Renovated Dormitory at the University of Rhode Island

Study by: REDE—Research and Design Institute, Providence Rhode Island, nonprofit, multidisciplinary agency

Methods used: Questionnaires and semantic differential questionnaires, observations, and interviews

Type of project: A before-and-after study as well as a comparative study (to an identical nonrenovated building)

Information source: REDE, *Butterfield Hall Evaluation Report*, 1974

INTRODUCTION

REDE is a nonprofit, multidisciplinary agency committed to investigating human behavior and developing physical environments that will more effectively meet the human needs of their users. Evaluations are carried out of its own design work which highlight both negative and positive aspects of the work. It is hoped that these evaluations will have a demonstrable effect on future designs. In the case of the evaluation of Butterfield Hall, for example, many of the residents' comments were negative in character, but REDE plans to use the information in future dormitory work.

Butterfield Hall and an identical dormitory—Bressler Hall—at the University of Rhode Island were considered poor places to live by the students as well as by the university administration. REDE was originally commissioned, in 1972, to remodel both buildings. The university's inability to raise matching funds from HUD led to a decision to renovate Butterfield only, on a greatly reduced budget. REDE's design goals included the incorporation of as large a variety of options as possible, a flexible design scheme permitting change and modification, the development of economical amenities to make dormitory life more attractive to students, and high quality at as low a cost as possible.

Construction delays put off completion of the project until spring 1974, and the evaluation was carried out during that time.

The Design Activity

Butterfield Hall is a four-story, boxlike, brick-faced building built in the late 1940s. It had housed male freshman students primarily. The renovation has lowered the percentage of upperclassmen leaving the dorm, but REDE emphasizes that there is no "average" resident and that their goal was to design the dormitory to be able to accommodate a diversity of behaviors from sleeping, to blaring stereos, to studying—all at the same time.

Meetings were held, open to the entire Butterfield community, observations were made, and interviews took place. The resulting design decisions included a variety of room arrangements, an activities corridor, a flexible furnishings system, and a geodesic roof structure. The latter was to provide space for recreation such as ball playing (taking place in the hallways before renovation). Although students felt that the roof structure was going to be the most exciting part of the project, budget cuts forced the elimination of that aspect of the job with the students' perception of having been "betrayed" as a result. The room designs resulted in two-man and four-man rooms, some horizontal connections and some vertical ones (connected by an internal spiral staircase). Some rooms were "across the hall," with a sleeping room on one side of the corridor and a study/social room across from it. This was an open study scheme in which the sleeping room could be closed off, or the moveable partition to the corridor could be removed, resulting in a semipublic space.

Carpeting was used extensively, sometimes on wall surfaces, but the budget did not allow for carpeting the individual rooms. By converting the room across the hall from the bathroom on every floor into a kitchen/lounge, the previously claustrophobic corridors became enlarged activities corridors.

REDE developed a furniture system in conjunction with the William Bloom Company of East Providence, which consisted of a free-standing wardrobe with shelves (for each student), a sleepscreen to attach to his bed (rejected by students and designers during evaluation), a hang-on worktop, a hang-on storage cabinet, two shelves (deemed insufficient for books by many students), three slotted uprights (wall tracks for shelving), and an adjustable chair.

The final budget approximated $250,000, with $180,000 for construction and $70,000 for furniture. A number of planned design features had to be eliminated, such as rigid acoustical ceilings, drawer storage, and picture tracks, in addition to the roof structure and carpet in the individual rooms.

THE EVALUATION

Although the REDE staff evaluated the progress of the renovation since students moved in during September, 1973, the systematic evaluation began after spring vacation, 1974, and included:

1. A questionnaire relating to general satisfaction with the dorm (distributed to all residents—135 out of a possible 173 responded).
2. A semantic-differential scale developed by David Canter administered to a random sample of 50 residents (40 responded) for purposes of comparison with a similar sample drawn before the renovation. A similar sample was attempted at Bressler (the non-renovated identical dorm), but few of the questionnaires were returned.
3. The University Residence Environment Scale (URES) developed by R. H. Moos of Stanford and M. S. Gerst of the University of California. The URES was administered to the same samples as the semantic differential scales.
4. Observations under the behavior mapping method.
5. Interviews with individuals and groups of students.

The report arranges the results so that the general quality of life in Butterfield is dealt with in one section and specific reactions to individual design elements in another. The URES and the semantic-differential as well as certain items in the general satisfaction questionnaire were designed to measure the residents' attitudes toward Butterfield as a place to live.

First, students were administered a series of semantic-differential scales (which require rating a concept on a series of bipolar dimensions such as "good–bad," "worst–best," etc.) involving an overall assessment of Butterfield as well as an assessment of specific factors (e.g., as a place to study, socialize, and sleep). For purposes of comparison, Butterfield residents had filled out the same scale prior to the renovation. In addition, a sample was also taken this year at Bressler Hall, a dormitory of identical construction in Butterfield that was not renovated.

Second, the short form of the University Residence Environment Scale was filled out by Butterfield and Bressler residents. The URES consists of forty true/false statements about dorm living which are scored for scales such as Involvement (in dorm activities), Support (residents give each other) and so on.

Based on the semantic differential scale results, with ratings from 1 to 7, it would be difficult to conclude that REDE has made a significant impact on the dormitory. Also, REDE's report states that other people in the dorm clearly exert a considerably stronger influence on satisfaction than does architecture. Figure 9 indicates that the dorm was originally, and still is after renovation, seen as a pretty good place to live. Similar figures based on the semantic differential questionnaire were prepared relating to the dormitory as a place to study, as a place to sleep, and as a place to socialize. In each case the pattern of the graph was quite similar to the one shown here.

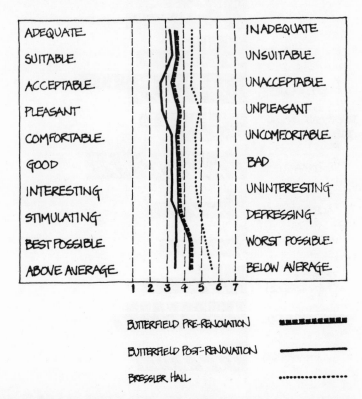

FIGURE 9. Student reaction to Butterfield Hall pre- and postrenovation compared to reaction to Bressler Hall.

The (URES) results are reproduced (Figure 10), and a copy of the table showing subscale descriptions is shown (Table 4). It must be pointed out that the graph for Bressler Hall is based on a very small sample since most of the questionnaires were lost to the researchers. Butterfield residents express attitudes not dissimilar from the national norm.

The preference for room designs, based on the general satisfaction questionnaire, was most favorable toward the double arrangement. The

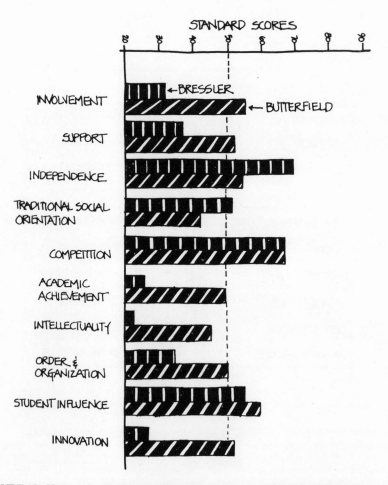

FIGURE 10. University Residence Environment Scale (URES) Profiles for Butterfield and Bressler Halls. (The national average on each subscale is 50.)

open-study and across-the-hall designs were least successful in the ratings and must be considered failures.

Residents of the dorm were asked to rate specific design elements on a scale of 1 (very successful) to 5 (very unsuccessful). Space was also provided for comments, and many took the time to write helpful statements. Students were consistently critical about the lack of drawer space in any part of the furniture system. One student commented: "I can't believe that in all of the research and 'insight' put into the renovation that such an elementary thing as a bureau or drawer space could have been overlooked." REDE's report also notes that "every student interviewed indicated a preference for the conventional grouping, most noting the lack of storage space plaguing the present system. When the question was rephrased so the wall-hung system would include adequate storage units, preferences were about evenly divided, some choosing the system approach for its flexibility and modern lines, others opting for the more traditional designs for their warmth and comparative durability." Typical comments ranged from: "This type of room is much easier to arrange as compared with the ones at Bressler" to "I like the more traditional concepts of furniture."

TABLE 4
Brief URES Subscale Descriptions

1. *Involvement:* Degree of commitment to the house and residents; amount of interaction and feeling of friendship in the house.
2. *Emotional support:* Extent of manifest concern for others in the house; efforts to aid one another with academic and personal problems; emphasis on open and honest communication.
3. *Independence:* Diversity of residents' behavior allowed without social sanctions, versus socially proper and conformist behavior.
4. *Traditional social orientation:* Stress on dating, going to parties, and other "traditional" heterosexual interactions.
5. *Competition:* The degree to which a wide variety of activities such as dating, grades, etc., are cast into a competitive framework.
6. *Academic achievement:* Extent to which strictly classroom and academic accomplishments and concerns are prominent in the house.
7. *Intellectuality:* Emphasis on cultural, artistic, and other scholarly intellectual activities in the house, as distinguished from strictly classroom achievements.
8. *Order and organization:* Amount of formal structure or organization (e.g., rules, schedules, following established procedures, etc.) in the house; neatness.
9. *Student influence:* Extent to which student residents (not staff or administration) perceive they control the running of the house; formulate and enforce the rules, control use of money, selection of staff, food, roommates, policies, etc.
10. *Innovation:* Organizational and individual spontaneity of behaviors and ideas; number and variety of activities; new activities.

Following are some examples of the ratings as well as some excerpts of comments:

Desk top   rating 2.45
VERY SUCCESSFUL   1 2 • 3 4 5   VERY UNSUCCESSFUL
Comments from "great idea—saves space, but poor material, as mine collapsed" to "full desk would be more useful."

Chair   rating 2.89
VERY SUCCESSFUL   1 2 • 3 4 5   VERY UNSUCCESSFUL
Comments from "conducive to study" to "terrible—back keeps slipping down; squeaks."

Wardrobe   rating 3.22
VERY SUCCESSFUL   1 2 3• 4 5   VERY UNSUCCESSFUL
REDE comments that the wardrobe works well for hanging clothes but is inadequate as the only piece available for storage. It was hard to move because of the undersized glides and there were problems with scratched floors (to be corrected the the installation of proper glides). Student comments included "the most successful piece of furniture" and "whatever happened to good-looking wood furniture?"

Blinds   rating 1.87
VERY SUCCESSFUL   1 •2 3 4 5   VERY UNSUCCESSFUL
Comments from "attractive" and "much better than curtains" to "I prefer curtains."

Floor Tile   rating 3.09
VERY SUCCESSFUL   1 2 3• 4 5   VERY UNSUCCESSFUL
Generally rated as adequate; however, many stated preference for carpet, and others noted that it was difficult to keep tiles clean.

The foregoing design elements were in the students' rooms; several examples of public spaces follow:

Booths   rating 1.80 (in kitchen/lounge area)
VERY SUCCESSFUL   1 •2 3 4 5   VERY UNSUCCESSFUL
"Good place to do homework, play cards, or just sit and talk" represented a more typical positive comment, although one student compared the booths to McDonald's and expressed dislike.

Seating   rating 1.99 (in main lounge)
VERY SUCCESSFUL   1 •2 3 4 5   VERY UNSUCCESSFUL
Most students seemed to like the appearance and the comfortable ambience, but some of the furniture was rated as being uncomfortable.

Blinds   rating 3.13 (in main lounge)

VERY SUCCESSFUL   1 2 3· 4 5   VERY UNSUCCESSFUL

"All in all, somewhat of a disaster. Slats are too fragile for the treatment received from both Mother Nature and the students. Once down, they tend not to be replaced." (comment made in REDE report)

The REDE report contains several recommendations with brief discussions. They are not intended as a comprehensive program for dormitory design but are viewed as an ongoing process of "hypothesize/design/build/test hypotheses/modify hypotheses." The propositions are:

1. Primary emphasis should be placed on study facilities.
2. All products, hardware, and material must be very "forgiving."
3. Personalization of public and private living spaces should be encouraged.
4. Careful attention must be paid to how students live in their spaces.
5. Storage needs are high for college students.
6. Security is a big problem in dormitories.
7. Flexibility is desirable wherever possible.
8. Privacy of the individual student should be respected.

A list of 20 specific recommendations (introduced as 20/20 hindsight) includes the reduction of density to two-man rooms and single rooms, require stereo earphones on all weekdays and nights, create a lockable drawer unit to hang off wall track, provide desk lamps, and carpet all corridor walls.

Lastly there is a modest list of 11 successful features which include two-man rooms, vertical suites, bathrooms, carpet in halls, window blinds, and dimmer switches. Some 70 pages of appendices contain seven sections as follows:

Corridor layouts

Basic room designs

Bloom furniture components

Renovation progress notes (these are mostly monthly reports from the evaluator to the architect. These observations in part have apparently been considered by REDE designers for "in progress" modifications.)

Evaluation materials (some are reproduced here)

Articles from the "Good 5¢ Cigar" (campus paper) relating to the renovation of Butterfield Hall

Original HUD proposal of 8.11.72 (that proposal sought funding support for Butterfield and Bressler dormitories)

DISCUSSION

Butterfield Hall Evaluation is a particularly thorough study. Several methods have been used to assure accuracy of data. Although the study was carried out by the same group who acted as designers, it did have the input of a researcher and assistant (William Aazano and Barbara Voye) in addition to the designers (Howard Yarme and Peter Wooding). Since the intent of the evaluation was geared toward ongoing improvement of design by REDE, it is particularly commendable and serves well as a good prototype for other design firms.

It is probably impossible to be completely objective in evaluating your own design, and ideally at least one outside researcher might have helped to establish totally objective criteria. The REDE report is honest and fair in reporting many negative data. In spite of many efforts, some students did not rate the design very highly and had many specific critical comments. Perhaps the key to the students' dissatisfaction was in their perception of REDE having taken over "their" dormitory and made it into REDE's.

REDE blames the budget cut for much of the negative student reaction. Particularly, what they feel was the most exciting part of their design—the roof structure—was eliminated, and students perceived that as a kind of betrayal.

Perhaps the total design should have been revised in view of the budget cuts, rather than the piecemeal elimination of certain parts that took place, while the original plans were clung to. It is very doubtful, however, whether economic reality would have permitted a totally new design proposal.

Students were involved in the original design process but did not perceive this involvement as meaningful. Hence it appears as if the students' participation was insufficient. Especially when dealing with students, an ongoing involvement seems very important in order to create understanding and positive attitudes toward the new design. From the students' point of view the designers came around to ask some questions before and after the renovation, but the real design decisions were clearly made by the experts.

In spite of some student criticism, there were many positive comments, especially when Butterfield Hall was compared to other dormitories. Above all, as a post occupancy evaluation study it is a well structured and thorough example of the evaluation of an interior renovation.

Some Major Causes of Congestion in Subway Stations

Study by: Gary H. Winkel and D. Geoffrey Hayward

Methods used: Observation

Type of project: An investigation of pedestrian movement in New York City subway stations as a predesign evaluation

Information source: *Some Major Causes of Congestion in Subway Stations,* 1971

INTRODUCTION

This study investigated pedestrian movement in several New York City subway stations. The study was initiated by the Urban Design Group of the New York City Planning Commission, whose primary concern was centered around specific elements and configurations of subways in need of renovation. The study was undertaken by a group of researchers in the Environmental Psychology Program of the City University of New York. The investigators, assisted by several graduate students, decided to focus on stairs and escalators, seating on platforms, entries and exits, and types of platforms. Each of these variables was studied in terms of its effect upon pedestrian flow.

The study dealt only with two-track subway stations during peak-flow rush hours. The project was an open-ended one, since it was intended to be a pilot study from which further research and demonstration projects would be conducted to verify the initial findings. Because of its preliminary nature, only observational techniques were used; in spite of this, the 83-page report contains some specific proposals for alleviating some congested conditions at the particular stations that had been investigated. Summaries at the end of each chapter also contain specific suggestions addressed to designers of new stations or to those who work on the renovation of existing stations.

The report is organized into five major sections: uniform work hours, similarity of platform entrance points, issues of stairs or escalators, connections to destination points, and issues of intrastation pedestrian movement.

STUDY DESIGN AND METHODS

The researchers made their preliminary decisions about what to study based upon previous reports and studies of subway station design by other researchers. The majority of stations selected were in Manhattan; one station was in Queens. Figure 11 illustrates a typical station layout. The observational methods used were tracking, behavior mapping, and counting. For tracking data, randomly selected persons were followed when entering a station. For distribution studies, measures were taken of where people wait for a train on particular platforms. Figures 12 and 13 show examples of the data compiled from observations.

Other aspects of pedestrian movements that were studied dealt with the capacity of stairs and escalators, the effect of column locations on platforms, the effect of bench and vending machine locations, reservoir spaces (near the stairs and escalators), and connections to destination points. Due to the preliminary nature of the project and the expectations of continued research, the main thrust of the study consisted of astute and structured observations, resulting in clearly stated comments and narratives.

Setting

The New York City subway system is large and complex. Although the focus of this study was limited to the two-track subway stations, mostly in Manhattan, many stations in the system provide transfer from

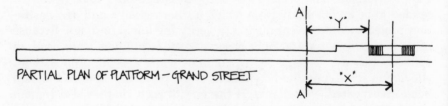

PARTIAL PLAN OF PLATFORM – GRAND STREET

FIGURE 11. Typical platform plan of subway station.

	Destination							
	Manhattan				Brooklyn			
	Morning		Evening		Morning		Evening	
	Rush	Other	Rush	Other	Rush	Other	Rush	Other
Median of where men boarded	68.'	139.'	49.'	63.5	87.'	145.'	64.'	83.'
Median of where men waited	67.'	129.5'	52.'	65.5'	74.'	139.5'	64.'	68.'
Median of where women boarded	77.'	34.'	—[a]	39.5'	69.'	112.5'	55.'	59.'
Median of where women waited	77.'	34.'	—[a]	46.'	97.'	102.'	46.'	62.'

[a] Too few observations.

Conclusion: There is no difference between where a person waits for a train, and where he boards that train.

Distribution studies are necessarily conservative in estimating the crowding of subway cars around the entrance points on a platform.

Our tracking study reported three different situations relevant to a person's arrival on the platform:

Situation No. 1: There was no train in the station nor coming in when the person
(173 people) arrived on the platform, nor did a train come in while the person was walking along the platform (this situation is the only situation which was recorded in distribution studies).

Situation No. 2: There was no train in the station nor coming in when the person
(62 people) arrived on the platform but a train came in while the person was walking along the platform.

Situation No. 3: There was a train in the station or coming in when the person
(52 people) arrived on the platform.

Because a distribution study would not have accounted for situation No. 2 or No. 3, and because those people boarded near the entrance point (especially situation No. 3 where a train was already in the station), it is apparent that the entrance peaks—as estimated by distribution studies—do not account for people who did not *wait* for a train, and therefore those estimates of entrance clusters are conservative.

FIGURE 12. Sample study results.

local to express trains in four-track stations, and many other stations have design features quite different from the ones that were studied here. Concerns such as noise level, air movement, lighting level, materials, frequency of train departure, etc., were not considered in context with this project.

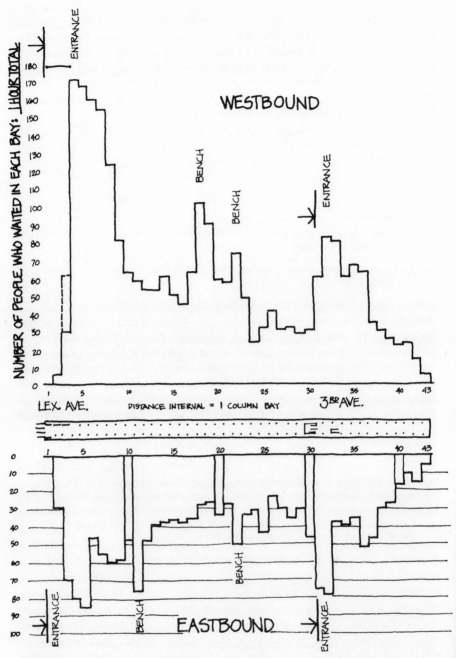

FIGURE 13. Distribution of people waiting for a train at the 54th St. and Lexington/3rd Ave. station.

Congestion occurs in typical New York subway stations during the peak rush hour periods—the beginning and the end of the workday, centered around the hours of 9:00 and 5:00. Some major employers have staggered work hours in order to lessen the peak load in transit facilities. This has alleviated some congestion in subway cars but has not reduced congestion within the particular station serving the major employer. It has simply moved rush hour conditions ahead in time.

The researchers made it clear that some congestion will always be present. There is a limit as to how many stairs and escalators the system can support. There is the further consideration of security during the non-rush-hours in station with many exits and entrances.

EVALUATION

It was observed that people tend to cluster around the bottom of stairs when entering a station. The next major grouping of people was around benches. It was thought that the clustering near stairs is because people tend to anticipate their destination points; due to the similarity of the entrance locations at other stations, most people prefer to board trains near the stairs. Two possible solutions for alleviating congestion due to clustering are suggested. The first is to entice people to even out their distribution through use of platform improvements such as better lighting, more benches, vending machines, and graphics. The second approach is to accommodate clustering at the entrance and exit points by providing more space at that point for people to stand. The possibility of alternating entrance locations along the route is also suggested.

The capacity of stairs and escalators is of major importance when examining causes of congestion. The researchers recorded the length of time people spend waiting to get up the escalators during the morning rush to be up to or even longer than 2½ minutes. The reverse flow of stairs is another serious circulation problem, as passengers try to enter stairs in the opposite direction of the crowds rushing up or down the same stair. Congestion can also be caused when the direction of the stair or escalator contradicts pedestrian flow. It has been observed that changes in direction, such as 90-degree turns, tend to slow people down (see Figures 14 and 15). Studies of walking speeds evidence that the change in those speeds due to a change of direction or a stair approach must be absorbed by the flow of people, or congestion will occur with even a light volume of people. Wider stairs and more stairs and escalators are suggested as one solution. A second possible solution is to provide greater walking distance from the edge of the platform to the stair or escalator. This greater distance will separate

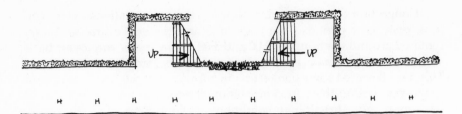

FIGURE 14. Plan of existing subway station.

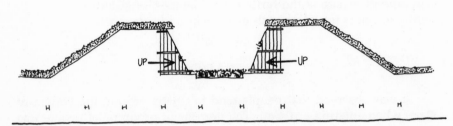

FIGURE 15. Proposed station plan to facilitate movement to and from stairways.

people by their natural walking speeds so that those who go more slowly will be last to reach the stair, not impeding the progress of those who walk more quickly.

Observations show that when stairs and escalators are not large enough to handle everybody, crowds form in front of the entrances and block circulation (see Figure 16). Reservoirs of space at the bottom of stairs and escalators are suggested as a solution to that problem. Sometimes the location of a stair or escalator in the middle of a platform creates congestion, since the space on either side is too narrow for passage, a condition frequently found on central platforms. For future platform design either platforms should be made wider, or stairs should be placed at the ends of platforms. Another observation points to the fact that many passengers are headed for a particular office building when exiting during rush hours; hence, direct access to more of the major buildings would help alleviate congestion.

It was noted that columns may actually have a positive effect upon platform circulation. People are more stable when they are stationary; therefore, on a narrow platform it is less dangerous to have those people standing to be along the edge, and those walking to be along the wall. Passengers tend to lean against columns rather than walls, especially when the walls are dirty and when advertising or vending machines are against the walls, and it is suggested that this pattern be reinforced on

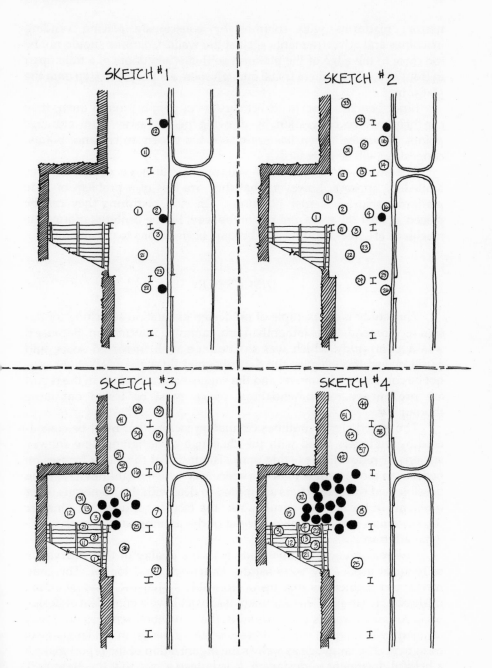

FIGURE 16. Four moments in the unloading sequence of a subway train.

narrow platforms with columns by consciously placing vending machines and advertisements against the walls. Columns should not be too close to the edge of the platform so that when doors of a train open in front of a column there is still enough room for people to step onto the platform.

Since benches tend to attract clusters of people around them, they can help relieve congestion by drawing people away from entrance points. However, if benches are placed too closely to entrance points, congestion will be increased.

The study of the effects of vending machines on circulation was limited. It appears, however, that they are less of a problem on side platforms than on center platforms. On side platforms they can be placed against the walls and out of the way, but on center platforms the machines are free-standing and become obstructions to circulation.

DISCUSSION

This study is an example of predesign evaluation searching for design recommendations rather than programming information. Because it was a pilot study which was to continue with increased scope and additional methodologies, the data obtained through observations cannot be considered definitive, and the suggestions contained in the report are preliminary recommendations which must be tested out more thoroughly.

The report clearly outlines circulation factors that must be considered by designers faced with the challenge of designing new subway stations or renovating existing ones. The fact that the study focuses on pedestrian movement results in a specificity which might not have been possible had the researchers attempted to deal with the numerous other problems in the subway system. Yet, the fact that it does not consider the system as a whole makes several of the assumptions appear somewhat arbitrary.

The real strength of this study is in the quality of the observation techniques used: they were astute, structured, and logical. The comments and deductions that result from this systematic observation are misleadingly simple. These trained observers have a good deal of design background as well as background in the behavioral sciences, and they came up with a report that addresses issues of design in the language of designers. The language as well as the organization of the report make it a helpful document to designers. It is indeed a pity that the New York Transit Authority did not lend its support to further investigations or demonstration projects along the same lines. However, later discussion with people in the City Planning Commission revealed that the preliminary station designs (general prototypes for the Second Avenue line) changed significantly after this research was submitted. The impression of the planners was that more stairs were added, stairway orientations were changed, wider platforms were recommended, and other conclusions from this report were also being considered.

Cambridge Hospital Social Service Offices

Study by:	Janet E. Reizenstein, Kim R. Spencer, and William A. McBride
Methods used:	Observation, interviews, questionnaire, photography
Type of project:	Predesign programming, design and Postoccupancy evaluation
Information source:	*Social Research and Design: Cambridge Hospital Social Services Offices*, 1976

INTRODUCTION

This study was conducted by three researchers/designers from Harvard's Graduate School of Design, in response to a request for assistance by the Director of the Social Services Department of the Cambridge Hospital, Cambridge, Massachusetts. The researchers had the unique opportunity to demonstrate the benefits of including social research in all stages of the design process. They proceeded from predesign programming through design and construction to postoccupancy evaluation. Describing in detail the process followed by the researchers, this study includes comparisons of the old and new offices and explains the research methods that were used.

STUDY DESIGN AND METHODS

New offices were required for a small group of approximately 20 social service staff members at the Cambridge Hospital. Although the offices are not unique, the thorough process of predesign programming, followed by design, construction, use, and postoccupancy evaluation by the same group of researchers/designers is an unusual endeavor.

The programming research was conducted through interviews with most staff members and was backed up with observations to understand

TABLE 5
Architectural Program

Type of space	Activities	Physical requirements
Social work (SW) offices		
General needs	Counseling and interviewing patients, families	Soundproofing, floor-to-ceiling partitions
	Meeting with hospital staff, agency personnel	Visual privacy
	Paperwork	Temperature control
	Telephoning	Lockable offices
	Talking informally	Amenities, including fresh
	Retreat	paint (neutral color), windows, carpet
		Furniture, including new desks with file drawers, comfortable chairs, shelves, bulletin board
Specific needs		
Psychiatric SWs	Interviewing and meetings with large groups (up to 10)	4 offices
		Location in Social Service Department
	Treating children	Room with one-way mirror
Medical/Surgical SWs	Interviewing, counseling small groups (up to 4)	5 offices
		2 located in Social Service Department, 3 on the wards[a]
Community SW	Meetings outside office and in office	1 office
	Health planning and elderly referral	Location in Social Service Department
Secretary's office	Answering phone	Same as for SW offices
	Office management	Proximity to Director and to SWs, but not right next to waiting area
	Typing	Filing cabinet area
	Monitors staff whereabouts	Space for volunteer
	Supervises volunteer	Blackboard
Director's office	Administration	Proximity to secretary
	Meetings with SWs hospital staff, agency personnel	Adjacent meeting room (to seat up to 15)
		Positive image
Student room	Same as for SWs	Large enough for 2 desks
		Amenities, if possible

(continued)

TABLE 5 (continued)

Type of space	Activities	Physical requirements
Meeting rooms	Staff meetings (up to 20) Large counseling groups (up to 10)	Large table and chairs Amenities
Waiting room	Clients and families waiting for SWs (up to 8)	Within sight of SW offices
		Comfortable Pleasant atmosphere Special considerations for elderly and children (chairs easy to get in and out of, play area)

^a If possible

existing patterns of behavior. Program research considered all physical needs, as well as the issue of image of the group in relation to the medical staff of the hospital. The findings were put into the form of an architectural program (Table 5).

Design and construction were limited by budgetary restraints, and not all recommendations were incorporated in the actual building of the offices.

Postoccupancy evaluation was conducted 4 months after the move into the new offices. Focused interviewing was the primary method of data collection. Those staff members who had been interviewed for programming purposes were interviewed again, and several clients as well as additional staff members were also interviewed. Observations were carried out over a period of several days. The third research method consisted of a short questionnaire which allowed quantitative comparisons between the old and the new offices. Finally, photography was used to document some physical features and use of the new complex.

Setting

Cambridge Hospital is housed in a building completed in 1968. The past and present Social Service Department offices are located in an older building, Cahill House, next to the hospital. The old offices were substandard. Two social workers, for instance, shared an office which had been a bathroom. Several members of the unit were in other locations of the hospital.

The new space assigned to the Social Service Department was still limited. It permitted uniting most members of the staff, but it necessitated the creation of several inside offices without windows.

Major Issues

Interviews done for predesign programming had revealed that users had special concerns with types of spaces, number of offices, size, privacy, social interaction, image, and design details. Through additional postoccupancy interviews and observations, the new space was compared to the old space. In spite of the fact that certain spaces could not be provided in the new complex (storage room, meeting room, children's waiting area), the evaluation of the new spaces is rather positive. Table 6 summarizes the comparisons.

Context of Social Service Department

Social service departments are non-revenue-producing and tend to be low in image and prestige in the social strata of a hospital. This partially explains the rather limited budget made available for the new offices. The jobs of social workers make privacy, both visual and acoustical, of great importance. Although clients at times prefer not to talk with their case workers and prefer conversations in areas which offer distractions, in general, both staff and clients consider privacy of great importance.

EVALUATION

The evaluation was focused upon interviews with the staff members. Questions were asked, and answers are reported, according to the categories of special concerns elicited from the staff members during the predesign programming (see under major issues above). Since the Social Services Department numbers only about 20 people, the results are presented in narrative form rather than in statistical or percentage figures. Many of the answers are given in form of direct quotations, such as the response of the director to a question about the new image of the department: "The environment makes a tremendous difference. I can see it. I can see what a difference it's made in the morale and self-image of my own staff to have a decent place to work." The observations conducted as part of the evaluation simply reinforced interview findings.

TABLE 6

Comparison of Old Offices, Design Program, and New Offices

★	Old offices	Design program	New offices
Types of spaces	SW offices, secretary's office	SW offices, secretary's area, Director's office waiting area, conference room, meeting room	SW Offices, secretary's area, Director's office, waiting area, family conference room
Number of offices	6 SW offices, 1 secretary's office	10 SW offices, 1 Director's office, 1 secretary's office	11 SW offices, 1 Director's office, 1 secretary's area
Relation of spaces	Director, Psychiatric SWs, waiting area separate from rest of dept., no offices on patient floors	All offices together. No offices on patient floors since none were available	Director, Psych. SWs, waiting area, secretary, conference room, SW offices clustered on Cahill 2.
Size	Medical/surgical SWs: 63 Psychiatric SWs: 208	Large enough to seat 4–5 people at once.	Outer offices: 83 Inner offices: 57

Privacy	Little acoustical privacy, some visual privacy	Acoustical and visual privacy are essential	Acoustical privacy, but some people complain about noise. Good visual privacy.
Social interaction	Occurred mostly in secretary's office.	Places to talk informally, place to retreat.	Centered in secretary's area, but also occurs in SW offices, hallway, and family conference room.
Image	Negative, depressing, dark, "The hospital doesn't think Social Service is important."	Positive, bright, cheerful, professional	Positive, bright, cheerful, "Social Service and its clients are important to the hospital."
Design details			
Furniture	Old, bad shape	New desks, chairs	Most same as before
Ventilation	Bad, fumes from parking lot	Important	Only OK if windows, doors open
Light	Little	Both natural and artificial	Plenty
Color	Varied, old paint	Neutral	Off-white, neutral
Temperature control	No AC radiators	Regulatable heat, AC	Some radiators not regulatable, no AC

TABLE 7
Office Space Questionnaire[a]

	Old office		New office	
1. What percentage of time is spent in your office?	29%	(average)	44%	(average)
2. Do other hospital personnel come to your office?	Yes	8(73%)	Yes	11(100%)
	No	3(27%)	No	0(0%)
3. How do you feel about having hospital	Bad	8(73%)	Bad	0(0%)
personnel to your office?	Neutral	1(9%)	Neutral	0(0%)
	Fine	2(18%)	Fine	11(100%)
4. Do you hold meetings in the Social Service	Yes	3(27%)	Yes	11(100%)
Department?	No	8(73%)	No	0(0%)
5. Is your office large enough?	Yes	4(36%)	Yes	9(82%)
	No	7(64%)	No	2(18%)
6. About how many clients do you see in your office each week?	5.2	(average)	10.0	(average)
7. a. About how many clients do you see on patient floors each week?	12.3	(average)	10.1	(average)
b. About how many clients do you see in the Outpatient Department each week?	7.1	(average)	7.7	(average)

[a]Sample size, = 11; Response rate, = 100%. (Note: Time percentages and number of clients are estimates.)

A brief questionnaire (13 questions) was administered to 11 staff members. The questions were designed to elicit comparative data between the old and the new offices. Positive responses far outweighed negative responses in answer to questions about the new spaces. A portion of the questionnaire is shown in Table 7.

DISCUSSION

The most significant aspect of this study is its basic simplicity. It was not meant to be a study requiring large sums of money, personnel, and sophisticated methods, nor was the design project itself an unusual one. It does, however, demonstrate clearly the benefits of including social research in all stages of the design process.

The fact that the research/design team had control of most steps from predesign programming to evaluation is rather unusual. It is, of course, an excellent example of the kind of evaluation that can easily be carried out by all design practitioners. The benefits of such an approach provide the input for future design projects. The process also has the additional benefit of involving the users in a real and constructive way from beginning to end. The following quotation from one of the staff members summarizes the success of the process:

> I thought the process was really a good one. I was very pleased with the way in which you came around and talked with everyone and got a sense of our jobs, and I really thought that that's really the only way to plan something for a group of people and that maybe what has happened in other places is that nobody thought to do that. It was also a very personal thing. I felt that you were getting to know us and you were creating something that was going to be useful for us.
>
> I was very impressed that you asked us what we wanted. That makes sense, and it seems too obvious that's what you should do, but in reality, most people never do that: never ask the people who are going to be affected by what they're doing what they want. They may ask them after it's all done, "How do you like it?" but not too often before it's done.

An Evaluation of the Fine Arts Center at the University of Massachusetts, Amherst

Study by:	Marion Brown, Tom Johnson, Yuji Kishimoto, Lynn Reynolds, Sumio Suzuki, Kathy Tepel
Methods used:	Interviews, archival research, questionnaires
Type of project:	A building evaluation based on research into its history and programming, and an evaluation based on nonusers as well as users
Information source:	*An Evaluation of the Fine Arts Center at the University of Massachusetts, Amherst*, 1975

INTRODUCTION

The Fine Arts Center at the University of Massachusetts, Amherst was formally opened in 1975. It is a nationally acclaimed building and probably one of the most significant works of architecture in the Northeast (Figures 17 and 18). A book dealing with the work of its architects, Kevin Roche and John Dinkeloo, features the Fine Arts Center on its cover, exemplifying the architects' pride in the structure. Almost every major architectural magazine has featured the building at one time or another.

It seemed of particular importance to begin a structured evaluation process of the building as early as possible, since like all important buildings, it provoked a good deal of controversy. An evaluation study was undertaken by six graduate students whose disciplines represented planning, architecture, interior design, and landscape architecture. The evaluation was initiated during a graduate seminar in design evaluation conducted by the authors.

FIGURE 17. View of the Fine Arts Center from the mall area (University of Massachusetts Photographic Service).

Similar to other institutional buildings, the Fine Arts Center went through a lengthy period of delays from the initial decision to build such a center in 1958 to its completion. The University administration had little experience in dealing with major buildings and none in dealing with major architects. Between 1958 and 1962, little progress was made and the first allocation of funds was made that year in the amount of $2,000,000. By that time, Pietro Belluschi had been retained as consultant to the university, and at his urging, a nationally known architect was retained. The first one to be asked was Minoru Yamasaki. Because of an insufficient allocation of funds for design fees, Yamasaki declined to undertake the commission and finally the job was given to Roche and Dinkeloo, with adequate fee allowances.

The study researched the project's history and programming. The study was conducted shortly after the completion of the building and prior to its full occupancy by all three departments slated as occupants. The questionnaires were analyzed using spss (i.e., Statistical Package for

FIGURE 18. View of the Fine Arts Center from the pond area (University of Massachusetts Photographic Services).

the Social Sciences, a commonly available "canned" computer program which allows statistical analyses to be completed with no prior programming experience).

STUDY DESIGN AND METHODS

The first part of the study dealt with the history of the building. Two of the researchers systematically went through records kept by the university, interviewed staff members, administrators, and faculty members who had been involved in the planning and programming process, and interviewed the architects for the building. Although a number of staff changes had taken place, most major "actors" were still available for interviews, and the various university departments were cooperative in making records and correspondence available to the researchers.

A second team of two researchers dealt with the nonuser campus population in an attempt to assess whether certain goals posed by the administration and trustees had been achieved (in the view of students and visitors to the University). A questionnaire (Figure 19, Table 8) was developed, using bipolar semantic questions and open-ended as well as yes–no questions, to gain an overall view of the individual's perception. One hundred eighty-nine questionnaires were completed by respondents, randomly selected near the campus pond (the Fine Arts Center adjoins the pond). The questionnaire elicited comparative responses to the two other major buildings around the campus pond, the new University Library and the Campus Center. The questionnaires were analyzed using SPSS subprograms CODEBOOK programming, CROSSTABS, as well as certain correlations using PEARSONCORR programs. Notice the percentages for distribution of perceptions on the following page.

The third team of two researchers polled the users of the building and studied their perceptions of the new structure. Responses pertaining to classroom space were not initiated since only three classrooms were in regular use at the time of the study. Approximately 200 students, staff, and faculty members responded to the questionnaires. (See Figure 20.)

Comments were written by over half of the respondents in both user and nonuser groups. Most comments dealt with suggestions for improvements.

Setting and Context

The Fine Arts Center is one of three major new buildings around the center of the campus, the campus pond. The other two buildings are the 29-story University Library designed by Edward Durrell Stone and the Campus Center designed by Marcel Breuer. Both of the latter buildings were initiated after the Fine Arts Center but completed before the opening of the Center. The Amherst campus of the University of Massachusetts is in an attractive rural area. The University serves 23,000 students, many of whom live in dormitories. The appearance of the campus resembles an urban center, somewhat unexpectedly seen in the context of the surrounding farmlands and hills.

The Fine Arts Center actually consists of seven buildings, housing the Departments of Art, Music, and Theater. (Speech was to be in the center but was eliminated years ago due to budget limitations.) It contains a large concert hall, a theater, and a smaller experimental theater, as well as all the major and minor spaces needed for the three depart-

A

Time: _____

Date: _____

Place: _____

CAMPUS BUILDING STUDY

The class of Environmental Design Evaluation at
the University of Massachusetts is conducting a
survey to assess whether several of the buildings
on campus are fulfilling some of the purposes it
was hoped that they would fill. The purpose of
this study is to make some constructive suggestions
for providing better buildings in the future. Your
cooperation in completing the following questions
will be most helpful.

B

Please indicate by a checkmark on the rating scale below how you would
describe each of the following buildings:

THE FINE ARTS BUILDING I

Impressive	( )	( )	( )	( )	( )	Unimpressive	/16
Beautiful	( )	( )	( )	( )	( )	Ugly	/17
Cold	( )	( )	( )	( )	( )	Warm	/18
Dynamic	( )	( )	( )	( )	( )	Static	/19
Inviting	( )	( )	( )	( )	( )	Uninviting	/20

THE LIBRARY I

Like	( )	( )	( )	( )	( )	Dislike	/21
In scale with the rest of campus	( )	( )	( )	( )	( )	Not in scale	/22
Functional	( )	( )	( )	( )	( )	Not functional	/23
Inviting	( )	( )	( )	( )	( )	Uninviting	/24
Dynamic	( )	( )	( )	( )	( )	Static	/25

THE CAMPUS CENTER I

Fits the site	( )	( )	( )	( )	( )	Does not fit the site	/26
A focal point	( )	( )	( )	( )	( )	Not a focal point	/27
Like	( )	( )	( )	( )	( )	Dislike	/28
In scale with the rest of campus	( )	( )	( )	( )	( )	Not in scale	/29
Functional	( )	( )	( )	( )	( )	Not functional	/30

THE FINE ARTS BUILDING II

Functional	( )	( )	( )	( )	( )	Not functional	/31
In scale with the rest of campus	( )	( )	( )	( )	( )	Not in scale	/32
Like	( )	( )	( )	( )	( )	Dislike	/33
A focal point	( )	( )	( )	( )	( )	Not a focal point	/34
Fits the site	( )	( )	( )	( )	( )	Does not fit the site	/35

CAMPUS CENTER II

Inviting	( )	( )	( )	( )	( )	Uninviting	/36
Dynamic	( )	( )	( )	( )	( )	Static	/37
Cold	( )	( )	( )	( )	( )	Warm	/38
Beautiful	( )	( )	( )	( )	( )	Ugly	/39
Impressive	( )	( )	( )	( )	( )	Unimpressive	/40

LIBRARY II

Cold	( )	( )	( )	( )	( )	Warm	/41
Beautiful	( )	( )	( )	( )	( )	Ugly	/42
Impressive	( )	( )	( )	( )	( )	Unimpressive	/43
Fits the site	( )	( )	( )	( )	( )	Does not fit the site	/44
A focal point	( )	( )	( )	( )	( )	Not a focal point	/45

1   2   3   4   5

FIGURE 19. Samples from questionnaire given to campus building users.

C

Regarding the FINE ARTS CENTER ALONE, would you please give an opinion
on the following:

1. Do you think this building is a good idea?   /46
    ( )very good   ( )good   ( )don't care( )poor   ( )very poor

2. Do you or would you like having a work area in this building?   /47
    ( )strongly yes ( )yes   ( )don't care( )rather not ( )definitely not

3. Would you bring your friends here to show off the campus?   /48
    ( )always   ( )usually ( )sometimes ( )seldom   ( )never

4. Do you go out of your way to go by this building?   /49
    ( )always   ( )usually ( )sometimes ( )seldom   ( )never

5. Do you like the material this building is made of?   /50
    ( )very much   ( )somewhat ( )indifferent( )dislike   ( )intense dislike

6. Can you think of this building as a peice of sculpture?   /51
    ( )very much   ( )somewhat ( )neutral   ( )rather not ( )definitely not

7. Do you find this building frustrating?   /52
    ( )very much   ( )somewhat ( )neutral   ( )rather not ( )definitely not

8. If you think this building has good and bad aspects, do the "good" /53
    aspects help you to overlook the "bad" aspects?
    ( )always   ( )usually ( )no opinion( )seldom   ( )never

9. Where is or where was your work area? _____   /54

10. Do you think of this building as symbolic of anything? ( ) yes   /55
    ( ) no If so, what?_____

11. What would you like to suggest for this building to improve it?   /57

    _____
    _____
    _____

Age_____   Sex M( ) F( )   /58
                                                                       /59
How many years have you been at the University?_____   /60
                                                                       /61
Occupation: Teaching Staff ( ) Office Staff ( ) Service Staff ( )   /62
             Student -- freshman( ) sophonore( ) junior( )   /63
                      senior( ) graduate( )

             Departmental Affiliation _____   /64

             .Visitor( )   /63
             .Other ( )   /63

FIGURE 19. (continued)

ments. However, a large part of the Art Department could not fit into
the new center. The building had been programmed and planned at a
time when the studio offerings at the University had been minimal. By
the date of completion of the building, the Art Department had grown
into a many-faceted unit, and only about half of the department's areas
are accommodated in the new building.

Although the program given to Roche and Dinkeloo was never a
complete statement of needs (one of the serious mistakes made by the

TABLE 8
Distribution of Perceptions

	Fine Arts Center[a]	Library	Campus Center
Impressive	77%	59%	39%
Beautiful	30	13	15
Cold	68	65	40
Dynamic	54	28	38
Inviting	30	19	47
Functional	35	41	65
In scale	30	18	57
Like	39	27	57
Focal point	51	68	65
Fits site	45	26	58
Good idea	68		
Like work area there	51		
Would use it to show off	43		
Go out of way to go by	9		
Like material	33		
Think it sculpture	66		
Find it frustrating	48		
Good outweighs bad	45		
Symbolic	48		
Not symbolic	52		
Positive symbol	20		
Neutral symbol	38		
Negative symbol	43		

[a] Percentages for first 2 columns are based on the SPSS CODEBOOK program.

client), several key points could be isolated from the original statement given to the architects:

1. Functionality and flexibility of the building for the expanding departments of Art, Theater, and Music.
2. "Front-door–back-door" focal point of campus (which has indeed been accomplished in both a symbolic and real way, as the building acts as a "gate" to the campus).
3. Human relationships to material.
4. Aesthetic influence on campus and community.

Design Activity

Roche and Dinkeloo were retained because of their national reputation and their experience in the design of fine arts buildings, theaters, music halls, etc. (When the firm was first suggested to the University, it was still headed by the late Eero Saarinen.) On the site allocated by the

University, the architects developed a highly sculpted exterior plan for a varied-height (four-story maximum), linked complex of buildings stretching out for over 600 feet. Their presentation of the preliminary designs to the University was heralded as "the most brilliant, professional architectural presentation that I have ever attended" by Richard Galehouse of Sasaki, Dawson, and Demay, who was giving some limited advice to the University on site planning and landscape at the time. The basic shape of the building changed very little from that initial presentation (Figure 21), and apparently little attention was given to the development of the interiors. It appears to be an approach in which function followed form. Due to budgetary limitations, changes had to be made such as the reduction of the concert hall from a capacity of 4000 to 2200. The building was meant to be a "monumental" one, and the idea of flexibility was obviously impossible. Expansion and change could not be achieved within the center's highly sculptural form.

Funds for the proper completion of the building—the interior design, graphics, identification, and landscaping—were never made available. Hence the perception and critical comments by many of the respondents in this study must be viewed in light of that reality.

EVALUATION

The archival research and the interviews resulted in several specific points brought into focus by the researchers as a kind of "lesson" for future buildings projects at the University.

1. The University, formerly an agriculturally and technically oriented one, was unsure about the real need of a Fine Arts Center, and the three involved departments were not prepared nor well enough organized to offer meaningful programs to the architects.
2. The rapid growth of the University added to the complexity of the program process. Despite surprisingly accurate predictions, the recent interest in art and other creative fields still turned out to be far greater than expected 15 years before the building's completion.
3. University administrators were inexperienced in dealing with a "monumental" kind of building. At least one staff member should have been retained who had prior experience with that kind of building.
4. The University accepted *verbal* promises given by one state administration and subsequently denied by the ensuing administration.

A   DATE _____

                                       TIME _____

                  PILOT STUDY: FINE ARTS CENTER

The goal of this study is to evaluate the Fine Arts Center in a

manner which will offer insight into user perception. This

questionnaire is a tool that will be used for the purposes of the

evaluation study and all forms will be kept anonymous. Your co-

operation in completing the following questions will be most

helpful.

AGE_____   ( ) MALE   ( ) FEMALE   58/
                                                                   59/
HOW MANY YEARS AT THE UNIVERSITY OF MASSACHUSETTS?_____   60/

( ) TEACHING STAFF   ( ) OFFICE STAFF   61/

( ) SERVICE STAFF   ( ) VISITOR

(.) STUDENT   ( ) FR   ( )SOPH ( ) JR ( ) SR ( ) GRAD

DEPARTMENTAL AFFILIATION _____   62/

FIGURE 20. Samples from questionnaire given to Fine Arts Center users.

---

  5. The University failed to prepare a real program for the architects
     and mistakenly expected the architects to solve all problems.
  6. While the Fine Arts Center was being planned and built, a whole
     new campus was being created with insufficient staff or outside
     consultants.
  7. The building was never completely finished due to insufficient
     funding. Hence, many negative user responses must be ex-
     pected.

Since monumentality and emphasis on exterior form were given
priority by the architects, it was especially crucial to examine the campus
community's perceptions of the design in terms of such descriptions as
impressive, dynamic, focal point, inviting, and *functional*. Ratings were indeed high for *impressive* (77%), *dynamic* (54%) and *focal point* (51%),

B

16. How would you describe the FINE ARTS CENTER?

impressive	( )	( )	( )	( )	( )	unimpressive	32/
beautiful	( )	( )	( )	( )	( )	ugly	33/
in scale with rest of campus	( )	( )	( )	( )	( )	not in scale	34/
warm	( )	( )	( )	( )	( )	cold	35/
dynamic	( )	( )	( )	( )	( )	static	36/
inviting	( )	( )	( )	( )	( )	uninviting	37/
functional	( )	( )	( )	( )	( )	not functional	38/
like	( )	( )	( )	( )	( )	dislike	39/
a focal point	( )	( )	( )	( )	( )	not a focal point	40/
fits the site	( )	( )	( )	( )	( )	does not fit the site	41/
	1	2	3	4	5		

17. If you have an office in the FINE ARTS CENTER, do you feel...?

proud	( )	( )	( )	( )	( )	embarrassed	42/
accessible	( )	( )	( )	( )	( )	inaccessible	43/
significant	( )	( )	( )	( )	( )	insignificant	44/
happy	( )	( )	( )	( )	( )	depressed	45/
comfortable	( )	( )	( )	( )	( )	uncomfortable	46/
safe	( )	( )	( )	( )	( )	unsafe	47/
effective	( )	( )	( )	( )	( )	ineffective	48/
private	( )	( )	( )	( )	( )	not private	49/
	1	2	3	4	5		

18. How would you describe the inside of the FINE ARTS CENTER?

fulfilled	( )	( )	( )	( )	( )	barren	50/
comfortable	( )	( )	( )	( )	( )	uncomfortable	51/
light	( )	( )	( )	( )	( )	dark	52/
organized	( )	( )	( )	( )	( )	disorganized	53/
stimulating	( )	( )	( )	( )	( )	dull	54/
attractive	( )	( )	( )	( )	( )	ugly	55/
appropriate	( )	( )	( )	( )	( )	inappropriate	56/
	1	2	3	4	5		

19. Do you have any additional comments? (e.g. suggestions) ·   57/

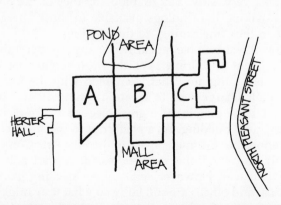

FIGURE 21. Simplified site plan of the Fine Arts Center: (A) Theater (B) Music Department (C) Art Department.

when compared with the other two buildings. *Inviting* rated only 30% and *functional*, 35%.

An identical group of questions was given to people who used the building on a day-to-day basis and to people who did not. Nonusers felt more positive about the building as *inviting, functional* and *liked*, while users felt more strongly on *focal* and *dynamic* (See Figure 22).

Faculty and staff among the users rated *privacy* and *happiness* rather low. This might have been due to the fact that all office doors had fully transparent glazing, and immediately after occupancy took place, every imaginable material appeared as a makeshift covering for doors. About 70% of the same group felt *comfortable*, while 61% of those users who did not have an office in the building responded that it is *uncomfortable*. Other negative responses reflected a dislike of the material (concrete) used inside and outside. Most users felt the interior of the complex to be *ugly* but *impressive*.

Many user comments were received in the categories of physical improvement, structural problems, and maintenance. The researchers felt that the many comments received were a positive indication that the building will continue to change and alter over a period of years.

DISCUSSION

There are two important aspects to this study which warrant special notice. One, the fact that a good deal of the research was concerned with issues that led the building into being—in other words, the history and the programming. The second important aspect of the study is the fact that it was done very early, before full occupancy of the building, with the intent of making the findings available to those charged with the completion and the "fine tuning" of the building.

The archival research has unearthed many disappointing problems caused by the client (the University), the most serious one having been the lack of a clear program. That, combined with the fact that the architects felt their charge was to create a "monument," brought into existence what many people call a great piece of architectural sculpture, but not a very good building. As a postscript to the research report, it is possible to report that the major space of the Fine Arts Center turned out to be a highly acclaimed and very successful concert hall. The theater space, too, works well. However, many of the teaching spaces, rehearsal rooms, studios, and offices are still far from what they might be, and are the ongoing subjects of unhappy complaints by the users.

Another major failure on the part of the designers is the lack of

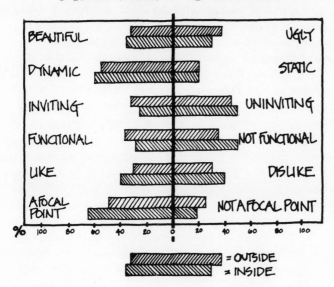

FIGURE 22. Responses to semantic differential questionnaires.

flexibility in the building, one of the clearly spelled out programmatic statements. The nature of the material (concrete) and the sculptural design make it almost impossible to consider any major changes or adaptations.

The evaluation by user and nonuser groups is in itself a very important way of obtaining a thorough set of variables in responses. Another study in this volume (Bryant Park) describes the involvement of two such groups. In the Fine Arts Center many of the quantified findings from a sample of close to 400 respondents were valid on those terms for serious consideration. The many specific suggestions made in response to the open-ended questions were also extremely helpful—or could have been. Unfortunately, none of the needed funds were made available to complete the building interiors, and certainly there were no funds for major improvements. But under normal conditions an evaluation conducted immediately upon the completion of a building can be extremely helpful to designers and architects in providing the finishing touches and in adjusting major problems. Especially in the case of the Fine Arts Center, this potential "humanization," based upon the findings of the research study, would have made all the difference to the success of the building, if it had been carried out.

In fairness concerning some of the critical comments about the building, one must, of course, keep in mind that any building (or any planned area, campus, neighborhood, project, etc.) needs time to age gracefully. The research report made it clear that the study was intended as only one of a number of evaluations to be conducted over the next few years. The chances are that even without the much needed funding for the completion of the Fine Arts Center, an evaluation conducted in several years would elicit somewhat more mellow and positive comments.

III

Buildings-as-Systems

INTRODUCTION

The previous chapter described several evaluations of interiors, focusing on the relationship of users to various interior design arrangements. The present chapter moves to a larger spatial scale. This chapter will scrutinize three evaluations which consider buildings as complete systems. Unlike Chapter II, these evaluations consider the functioning of entire buildings rather than simply of specified interior spaces. Yet, the focal problems of these evaluations tend to end at the skin of the buildings, while the studies in Chapter IV will extend the focal problem to include characteristics of the site.

The evaluations in this chapter represent a considerable diversity. The ELEMR (Effects of Living Environment on the Mentally Retarded) Project and the New England Villages study both evaluate housing for developmentally disabled people. The ELEMR Project is an extremely detailed 4-year analysis of a single setting. This project examines the impacts of renovations on developmentally disabled residents and direct-care staff at a large state institution. Using several techniques, such as direct observation, interview, and participant observation, the evaluators were able to document the influences of greater privacy and better living facilities for the users. In contrast, the New England Villages study was a quicker evaluation of a more normalized small "village" for developmentally disabled people. Using interviews, observation, analysis of documents, photography, and other methods, the evaluators examined the relationship of the buildings to larger social issues and assessed the effectiveness of individual spaces. In a study of a garden apartment complex, Zeisel and Griffin and a graduate design class compared the original intentions of the designers to the users' perceptions of the completed complex. The evaluators gathered data primarily through interviews with designers and residents.

In another case, Zube *et al.* evaluated and compared twelve National Park visitor interpretive centers. Using interviews, questionnaires, and direct observations, the evaluators judged the performance of the centers for users, staff, and administrators. The focal problem included both physical problems such as roof leakage and social concerns such as satisfaction with the interpretive programs of the centers. Although the four studies in this chapter are diverse, they contain several common methodological and conceptual themes. For example, they are focused on *users*; other aspects of the "structure for evaluation" are seen in relation to them. Moreover, all the evaluations are intended for *use*; each report contained many suggestions for designers, policy planners and users; most of the reports contained many graphics and illustrations. Finally, all of the studies used a combination of several methods so that the weakness of one method may be compensated by the other.

Effects of the Living Environment on the Mentally Retarded (ELEMR) Project

Study by: R. Christopher Knight, Craig M. Zimring, William H. Weitzer, Hollis C. Wheeler
Methods used: Interview, participant and nonparticipant observation, analysis of documents, unobtrusive measures, direct observation of behavior
Type of project: Longitudinal observation before and after renovation at a state training school
Information source: *Social Development and Normalized Institutional Settings: A Preliminary Research Report*, 1977

INTRODUCTION

The past 20 years have seen increasing pressure to integrate into society isolated groups such as the handicapped, racial minorities, and institutionalized populations. Spawned by this zeitgeist, the normalization principle (Wolfensburger, 1977) suggested important architectural and programmatic changes for institutions for people called "developmentally disabled."* This principle has been defined as: "making available to the mentally retarded patterns and conditions of everyday life which are as close as possible to the patterns of the mainstream of society" (Nirje, 1968). And, indeed, traditional institutions are almost totally antithetical to this principle. Residents of institutions often sleep in large wards; they eat in large dining halls; they aren't permitted to have sexual relations; they live in large poorly furnished buildings; they receive little training or stimulation.

The ELEMR Project is a longitudinal evaluation of institutional renovations which were guided by the normalization principle. The in-

*This term is the label currently preferred for groups previously labeled "mentally retarded."

stitution, a large state training facility, had a familiar institutional environment prior to the renovations: large 30' by 40' sleeping ward for 20–25 people, sparsely furnished tile-faced day halls, institutional social patterns, including poorly trained direct-care staff having the most contact with residents, and few training or recreational programs. After the renovations, the residents had a modicum of privacy, as well as new furniture, better lighting, and so on. Also, the social programs were somewhat improved.

The ELEMR Project was co-directed by one of the present authors (C.M. Zimring) with A. Friedmann as Principal Investigator, in collaboration with a multidisciplinary research team*. ELEMR serves as an example of an in-depth evaluation of a single setting. The professional research team spent nearly 4 years studying the institution and collected many thousands of pages of data. The conceptual scheme used in the evaluation serves as a good example of the SPA, however, and can serve as a model for less ambitious evaluations.

The focal problem in the ELEMR Project was to examine the transactions of the immediate community of residents, staff, and built environment. Attention was focused on these three elements, and the interrelations of interest were the transactions between them.

The larger system of influences included the other elements in the institutions, such as the administration, the professional staff, and the parents' association, and included outside influences, such as economic trends, the normalization principle, and state funding for new programs.

METHODS AND STUDY DESIGN

The ELEMR Project is a longitudinal (3+ years), multidisciplinary research project which is attempting to assess the impact of the built environment on mentally retarded people. Using a multiple-baseline quasi-experimental design, state-school residents and attendants were observed before and after their living quarters (large sleeping wards and day halls) were converted into smaller, more homelike spaces.

The designed environment was documented both before and after renovations. This included photo descriptions of all spaces, floor plans, narrative descriptions of interior materials, and an analysis of the acoustical quality of the buildings, including reverberation time and sound pressure homogeneity in various areas of a room.

*R. Christopher Knight, co-director; Harold Raush, Co-Principal Investigator; William Weitzer, Hollis Wheeler, and others, staff members.

Eighty randomly chosen residents were observed during several 6-week periods before and after the environments were normalized. Data were coded using a 40-item behavior checklist, and behavior was randomly sampled throughout the afternoon and evening. Time samples of behavior allowed assessment of proportion of time spent, for example, in social interaction, solitary behavior, stereotyped actions, or use of manipulable objects. In addition, each observed behavior was coded for its location in the building (4' × 4' grids) and any physical artifacts involved (e.g., chair, table, window, pole, T.V.).

A separate observation scheme was used to record both frequency and quality of staff–resident interactions. It was determined who initiated these interactions, and a variety of qualitative dimensions was documented: whether the initiation was a command or a question, apparent emotion, mode of communication (verbal or physical), and context (e.g., social, personal care, formal training). These interactions were followed through three steps: observing initiation, response, and any consequences to the response. As with the resident observations, these interactions were coded for location and physical artifact.

The project orientation dictated a broad methodological approach: a wide range of complementary observational methods were employed. Beyond the quantitative procedure described above, participant observation, staff interviews, unobtrusive measures (e.g., school records, breakage, building logs), and "critical incidents" were used.

The ELEMR Project structured much of its research around a "multibaseline quasi-experimental design." This design takes advantage of the staggered building schedule at the state school by permitting unrenovated buildings to serve as natural control groups for renovated buildings. This helped to control for general changes in the school treatment milieu. Also, whenever possible each individual is observed several times prior to and following renovations. This helps to separate the genuine effects of renovation from those simply brought about by novelty.

The Setting

The architectural environment* reflects the treatment model accepted during the State School's construction in the 1920s and 1930s. Most of the residents are housed in moderate-sized dormitories which

---

*This study deals principally with the fixed architectural environment. Although other aspects of the environment such as upkeep, furniture arrangements, and decoration are important, they are dealt with only peripherally as it focuses on more fixed characteristics of the built environment.

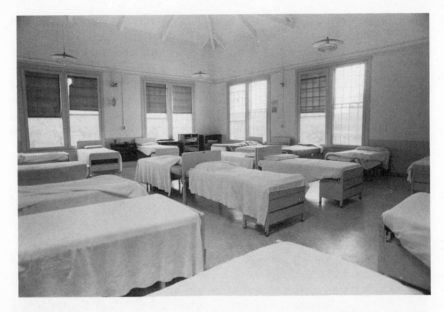

FIGURE 23. Typical ward at State School prior to renovations (Photograph courtesy of Alyce Kaprow).

are sited on a rolling campus.* At the outset of the ELEMR Project each contained six 30' × 40' spaces, three of which slept 15–20 residents in a ward arrangement. The remaining rooms served as day halls, dining rooms, or multipurpose rooms. These rooms were designed in the familiar institutional scheme, using asphalt tile floors, ceramic tile walls, and plaster ceilings, as illustrated in Figure 23. Furnishings were sparse, tended to be institutional in design, and were often in poor condition.

Many of these dormitories have undergone renovations and have been transformed into somewhat more homelike spaces, with modular units being introduced in some buildings, partitions in others, and full-wall construction in yet others (Figure 24). The modular units provide space for individuals, the partitions separate the large rooms into one- to four-person spaces, and the full-wall units have single or double bedrooms. The modular units have 4.5-foot-high walls which are intended to provide privacy while seated or lying down. The walls are joined together in "L" or "T" shapes and have built-in wardrobes, dressers, and work surfaces. The L- and T-shaped configurations are the basic units and are arranged to sleep 12–14 people in the large rooms, while also providing a common lounge space (Figure 25). The partitions

*There is presently an active effort underway to move many residents into appropriate residences in the community.

FIGURE 24. Renovated ward at State School using modular units for privacy (Photograph courtesy of Alyce Kaprow).

are 8 feet high and consist of sheetrock painted in a variety of colors. The partitions formed suite-type units, with 3 bedrooms for 2-4 persons, surrounding a common lounge space. In all arrangements the bathrooms are modernized with the addition of private toilet stalls and private showers.

The full-construction type of unit was a renovation of a more modern building. Although the building was completed in 1968 it had institutional features similar to its older counterparts: tile walls and floors, large reverberant spaces, and room for only few activities. After the renovations, the building resembled a college dormitory, with three floors consisting of two wings of doubled-loaded corridors of single or double bedrooms with central lounges.

Although the grounds were not part of the focal problem, they were quite notable. The 18 buildings which constituted the state school were sited on over 200 acres of rolling New England countryside, providing a seldom-used opportunity for outdoor activities.

Social—Historical Context

As was mentioned above, the focal problem had to be viewed in terms of the local and national economic trends and of changing treat-

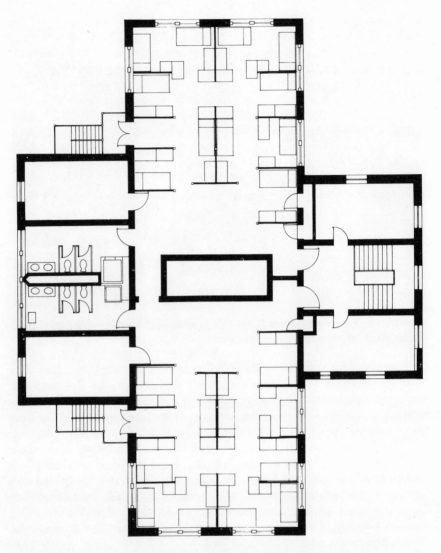

FIGURE 25. Floor plan of modular design. Bold lines within the large rooms are 4½-foot-high walls. Each module contains a bed, dresser and desk.

ment philosophy. The state school originated as a custodial facility and maintained many qualities of that genre. Resident–staff ratios were high, staff were poorly trained, and there were few programs. The direct-care staff had the most day-to-day contact with the residents, yet they were generally unskilled workers and made a wage similar to the production-line workers in local light industries. Professional employees were paid more poorly than their peers in the private sector and were attributed a low status. However, since the local economy was poor, an increasing number of well-qualified people were unable to find other jobs and were serving as direct-care and professional staff.

Despite these pressures toward poor care, however, the normalization principle was making increasing inroads into the institution. This principle dictated that the residents be given patterns as close as possible to the patterns of everyday life, and it had broad impacts such as influencing the class-action suit which resulted in the renovations, causing better, more enlightened administrators to be hired, bringing about increased staffing, and accelerating efforts to move residents into the community.

Proximate Environmental Context

The institution is set in a town of 5000 in a sparsely populated area in western New England. The nearest urban center, Hartford, Connecticut, is about 60 miles away. Also, there are a number of colleges and universities within 20 miles of the site. This proximity encourages students to enter volunteer programs while in school and to seek employment at the institution after they graduate.

The Users

The users in the ELEMR Project consisted of two groups: the developmentally disabled residents and the direct-care staff. There were 850 residents at the State School at the origin of the project in 1974, and by 1977 the number had been reduced to about 700, with plans to reduce the number to about 350 by 1980. These residents were all labeled severely or profoundly retarded, although these labels are quite unreliable at best. Indeed, the residents ranged in functional ability from self-sufficient, fully employed individuals to those who required assistance in the basic skills of eating, dressing, and toileting. Although the institution housed residents ranging in age from 7 to 80 years, the ELEMR Project focused on adults over 18 years old. In addition, the Project principally studied residents who had moderate to poor functional skills. Some policy planners doubted that this group, especially, would benefit from renovations. This was the focal problem.

The direct-care staff were a highly heterogeneous group. They ranged from elderly staff members who had been working 20 years or more with little training, to young graduates of the local colleges who saw the job as an entry to the health care field. All staff members, however, including the administration, were quite poorly paid in comparison to their counterparts in the private sector.

EVALUATION

The final report of the ELEMR Project was being prepared while this book was being written. A preliminary report, however, suggests several findings. First, the residents recognized and used private spaces when those spaces were well defined. All residents used their private spaces in the renovated dorms somewhat more than they did in the unrenovated spaces. Also, residents who moved into the full-wall college-dormitory-type arrangement used their private spaces more than did the residents in the modular arrangement. Second, when there was proper physical normalization, the renovations accompanied an increase in various socially valued activities on the part of residents, such as positive social behavior. For example, as illustrated in Figure 26, during the renovations positive social behavior increased from 6.8 to 10.9% for the one and two person bedroom design, yet remained about constant for the module design during the same period.

Yet the residents and the renovations cannot be considered without looking at the direct-care staff.

For purposes of illustration, consider the pattern of results in three renovated buildings shown in Table 9. In physical design, Gateway House* provides more clearly defined private spaces with full walls and doors that lock. In contrast, Beacon and Cardinal Houses are designed with single person modules, 4½-foot high walls, with bed, desk, and mirror in each space. So in physical design for privacy, Gateway appears more clearly defined than Beacon and Cardinal. Also, the social skills of the residents who moved into Gateway were more highly developed than those moving into Beacon and Cardinal, while residents in Beacon and Cardinal were comparable. Based on physical design for privacy and social skill levels to respond to these designs, one might expect residents of Gateway to recognize and use those spaces more readily than residents in Beacon and Cardinal. One would expect Beacon and Cardinal to be comparable.

The final report, which was published too late for detailed analysis in the present volume, revealed somewhat different findings. Although it is not possible to do justice to the 400 page report, a general finding was that the dormitory style renovation was the most effective one for even very low functioning residents. This environment was best because it offered the residents the greatest opportunity to control their personal experience by providing single and double bedrooms and a good hierarchy of spaces.

*These are fictional names of dormitories at the State School which were added to increase clarity.

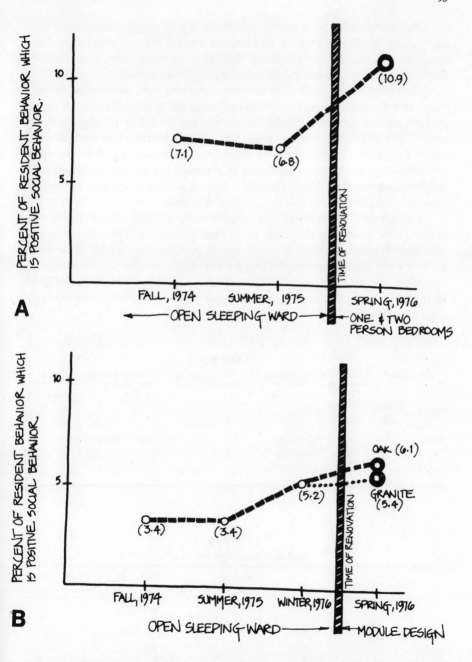

FIGURE 26. Comparison of behavioral reactions to open sleeping ward vs (A) one- and two-person bedrooms and (B) modular design.

As can be seen in Table 9, however, these expectations were not supported. Rather, Gateway and Beacon appear comparable while Cardinal shows much lower use by residents of their own space and a higher frequency of intrusion into other's private areas.

However, these data only consider the physical definitions of privacy. If we consider social definitions as well, the staff's use of these private areas must be considered. As we see in Table 10, the staff intruded into residents' private areas appreciably more in Cardinal than in either Gateway or Beacon. It is clarifying to understand that the staff intrusion data are clearly paralleled by resident intrusions. The staff and residents in Gateway and Beacon seemed to recognize the designed areas as private while those in Cardinal did not.

Participant observation and interviewing further clarified and expanded the meaning for data in Tables 9 and 10. These techniques revealed that a trend across buildings for residents to spend more time in their own private spaces during the evenings was due to staff encouragement and a drawn-out bedtime routine. This routine included getting night clothes on, staff directing residents to their private areas to get their clothes, etc.

TABLE 9
Residents: Time in Own Space and Intrusion into Others' Private Spaces

	Time of day	In own space (%)	Intrusion (%)
Gateway	Afternoon	86	14
	Evening	90	10
Beacon	Afternoon	No access to private spaces	
	Evening	84	16
Cardinal	Afternoon	3	97
	Evening	38	62

TABLE 10
Staff and Resident Intrusions

		Resident intrusions (%)	Staff intrusions (%)	Resident rank	Staff rank
Gateway	(Aft.)	14	5.7	2	2
	(Eve.)	10	1.2	1	1
Beacon	(Aft.)	No access to private spaces			
	(Eve.)	16	7.5	3	3
Cardinal	(Aft.)	97	16.5	5	5
	(Eve.)	62	15.8	4	4

The differential effects in Beacon and Cardinal buildings described in the tables could be traced to common and intense pressure on the staff from administrators and the parents' organization to perform supervisory and custodial roles. This pressure was effectively maintained through sporadic and unannounced inspections for safety, cleanliness, and neatness in the buildings. Beacon responded to this pressure toward cleanliness and order by keeping the residents in an open basement recreation room for easy supervision during the afternoon. The staff in Cardinal building allowed residents to use private areas but felt compelled to roam through these areas to check on safety and generally supervise behavior. This staff behavior, incidentally, demonstrated that designed spaces were not private but in fact routinely accessible. Those afternoon experiences seemed critical for residents' understanding of the definitions of designed spaces.

In the evenings when both Beacon and Cardinal staffs encouraged residents to use their private areas, the residents of Cardinal building had not learned to understand these spaces as their own, but rather had experienced clear demonstrations that they were not.

DISCUSSION

The ELEMR Project provides an example of a very complete analysis of a single setting. The principle focus of the study was theoretical, addressing the question: "How do people relate to their designed environment?" The multidisciplinary evaluation team had the advantage of a long time period to spend on the evaluation as well as considerable financial resources. Hence, a number of in-depth techniques could be used, including extensive direct observation. The longitudinal, multimethod approach allowed complex or seldomly occurring activities to be monitored by participant observation and interview, and allowed as well subtle trends to be detected through direct observation and quantification of behavior. Not many evaluations have the resources to scrutinize a setting with the completeness of the ELEMR Project; such scrutiny is needed, however, if we are to understand the relationship of peoples' activities to their designed setting.

Designing for Mentally Retarded People: A Social–Environmental Evaluation of New England Villages, Inc.

Study by:	Janet Reizenstein and William McBride
Methods used:	Participant and nonparticipant observation, interviews, analysis of documents, unobtrusive measures, photography
Type of project:	Resident and staff evaluation of a small sheltered village for the mentally retarded in Massachusetts
Information source:	Report to be published by the Architectural Research Laboratory, University of Michigan, Ann Arbor, Michigan, 1978

INTRODUCTION

For years mentally retarded people have been shut in harsh, dehumanizing institutions. There is an increasing realization, however, that people may only achieve their full potential if they are exposed to the patterns and expectations of everyday life. This realization prompted the renovations at a large state institution which were evaluated in the previous case. A number of critics have argued, however, that any living situation where people are housed in large institutions is inherently destructive of personal growth. The present evaluation examines a sheltered village where a small number of mentally retarded people live and work in a supportive setting.

This study is a careful and quite thorough "social–environmental" evaluation. It was paid for out-of-pocket and data gathering was done in a short period of time. Yet, a well-thought-out methodology and an articulated conceptual position helped make the study useful and comprehensible.

STUDY DESIGN AND METHODS

This study employed several methods, including focused interviews, participant and nonparticipant observation, attention to physical traces, photography, and examination of documents.

Interviews were conducted with three members of the board of directors, the Executive Director, the architect of record, and with residents and staff of the village.

The research team spent several days in each of the areas of the village: the sheltered workshop and the houses. They recorded their observations both while participating in meals and other activities and while passively observing.

The evaluators also observed the "physical traces" left by the users. These traces included wear-and-tear on the physical plant as well as signs of use of different areas.

Extensive photography was performed both as data collection and for illustration of the results in the final report. Records, correspondence, and other documents were examined to understand the design process.

The study design consisted of several visits to the site during a 5-month period from October, 1974 to February, 1975. The evaluators attempted to approach the initial open-ended interviews and review of documents with as little bias as possible. Information from these qualitative methods helped in establishing an operating heuristic for the remainder of the study. This heuristic consisted of three issues: (1) *social contact*, the degree to which the physical environment affects the amount and quality of social interaction, (2) *activity support*, the degree to which the physical environment allows a normal range of daily activities and facilitates successful accomplishment of tasks associated with these activities, and (3) *symbolic identification*, the information an environment conveys about the people associated with it. This scheme permitted the evaluators to assess the site, the house and the workshop with respect to broad social–environmental issues associated with the normalization principle: control (by residents), personal growth, and integration into society.

Setting

New England Villages, Inc., is a small settlement for mentally retarded adults on a 75-acre site near Pembroke, Massachusetts. The Village consists of three houses (each for eight retarded adults and two coordinators) and a large sheltered workshop. Twenty-four men and women lived at the village, and approximately 50 came in to work at the workshop during the day.

The site consists of the three clustered houses, with the sheltered workshop set close by. The houses are U-shaped with the "public" areas (living room, dining room, kitchen, entry, sitting room, guest bath, and laundry) at the center and two "private" areas (three double and four single bedrooms and 5 bathrooms) in the wings. The houses are built with wood construction and a pitched roof.

The workshop is a fairly large structure (about 5000 square feet) which is intended to feel like a New England meeting house. The plan of the workshop consists of a large central meeting space which is surrounded by several smaller rooms (e.g., kitchen/canteen, offices, toilets, storage, small workshops).

Design Activity

The Village was created by a group of parents, relatives, and friends of the mentally retarded in order to provide a supervised setting in which retarded adults "can live in dignity, with appropriate independence, work in jobs suited to their interests and capabilities, and enjoy leisure time with peers in a manner providing enjoyment and a sense of belonging."

A decision was made to build a small prototype village which would be attractive, modern, and intimate. Money was raised to purchase the site and a $216,000 federal construction grant was obtained to fund the structures. The architectural program was jointly established by the architect of record (M. Wyllis Bibbins) and the Village board.

The houses were to reflect a comfortable family setting similar to a house in which the architect would like to live himself. The program for the workshop was more complex and less clear. Vocational, administrative, and recreational functions had to co-occur in a climate which supported growth and independence. Also, whereas the architect and the Board had previous experience with houses, no one knew precisely what the workshop should be like.

Social–Historical Context

Several factors constituted the context for evaluation: (1) the social and policy context, including recent work in the study of mental retardation and resulting treatment philosophies, and (2) the immediate physical context, including the neighborhood. Mental retardation (or the more preferred term "developmental disability") is a complex phenomenon which can be caused by cognitive damage resulting from genetic causes, disease, malnourishment, or mistreatment. More subtly, however, the manifestations of mental retardation are influenced by labels that we establish for people who differ from the norm. If we view "retardates" as sick, subhuman, pitiful, or menacing, for example, we set up self-fulfilling expectations which encourage behaviors which fit these descriptions. The power of such labels influenced many professionals to adopt the normalization principle, defined as "letting the mentally retarded obtain an existence as close to the normal as possible." New England Villages stemmed from this principle. It attempts to set up a homelike environment which duplicates many of the patterns and rhythms of "normal" existence (e.g., a separation between work and home, the availability of truly private spaces).

Proximate Environmental Context

The physical context of New England Villages consists of its relationship with the town of Pembroke and with shopping and recreation facilities. For example, because the Village is not connected to community resources by public transit, residents must be driven to shopping and entertainment. The 75-acre site had potential for outdoor activities, yet the site was not developed at the time of evaluation.

Users

Twenty-five residents lived in the Village, eight in each of three houses. By traditional criteria these residents would have been labeled mildly or moderately retarded. Most of these residents worked in the workshop, although some had jobs in the community. In addition, about 50 developmentally disabled people came from the community to work in the workshop during the day.

EVALUATION

A five-stage model of the design process was chosen as the structure for analyzing the physical environment of New England Villages, Inc. Stages in this model include: predesign programming, design, construction, use, and evaluation. The physical environment was divided into site, house, and workshop. The house was analyzed as a whole and according to its component spaces. The workshop was also examined as a whole and according to its component spaces. See Figure 27 for

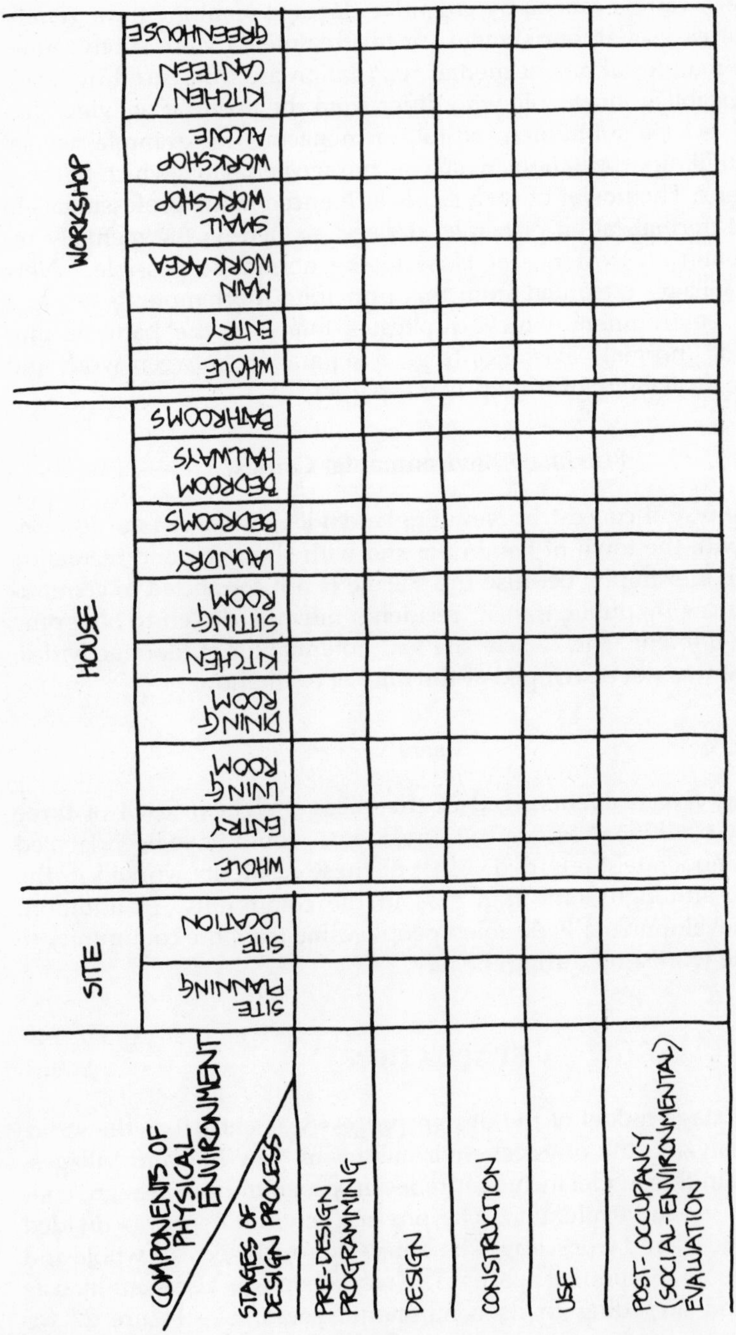

The following table structure is depicted in the figure:

COMPONENTS OF PHYSICAL ENVIRONMENT / STAGES OF DESIGN PROCESS	SITE		HOUSE										WORKSHOP						
	SITE PLANNING	SITE LOCATION	WHOLE	ENTRY	LIVING ROOM	DINING ROOM	KITCHEN	SITTING ROOM	LAUNDRY	BEDROOMS	BEDROOM HALLWAYS	BATHROOMS	WHOLE	ENTRY	MAIN WORK AREA	SMALL WORKSHOP	WORKSHOP ALCOVE	KITCHEN / CANTEEN	GREENHOUSE
PRE-DESIGN PROGRAMMING																			
DESIGN																			
CONSTRUCTION																			
USE																			
POST-OCCUPANCY (SOCIAL-ENVIRONMENTAL) EVALUATION																			

FIGURE 27. Structure for data analysis: interaction of design process model and physical environment.

the way the design process model and physical environment variables interact.

The evaluation was analyzed on four levels: (1) by detail, which included design variables (e.g., lighting), their related design decision (e.g., surface-mounted incandescent), and behavioral implications (e.g., not enough light for some activities) for each identifiable space; (2) evaluation by space, which describes what each space looks like, how it is used, and what the residents, staff, architect, board of directors, or research team think about it; (3) evaluation by issue (social contact, activity support, and symbolic identification); and (4) conclusions which summarize the degree to which the physical environment supports the normalization concerns of control, growth, and integration.

Specific suggestions to New England Villages, Inc., and general design guidelines for small residential settings and sheltered workshops for mildly and moderately mentally retarded people were formulated from research findings and information found in literature on the normalization principle and are presented at the end of the evaluation. For example, house design was evaluated by detail and by space as follows.

House Design

As stated earlier, the three houses are U-shaped with the "public" areas (living room, dining room, kitchen, entry, sitting room, guest bath, and laundry) at the center and two "private" areas (three double and four single bedrooms and bathrooms) in the wings (Figure 28). In addition to spaces called for in the program, a laundry room, sitting area, and two porches were included in the plan. The architect used wood construction and a pitched roof to achieve residential character and scale.

Construction

The only major change during construction was that porches designed for each house were not built, owing to cost considerations. There were to be two porches in each house: one off of the dining room and living room, and one (a deck) in the entry court.

House Evaluation

The houses are "homelike" in appearance, and their color and texture blend with the surrounding woods. The plan includes the components of a "normal" house with the exception of storage spaces such as a

TABLE 11
House Design

Design variable	Design decision	Behavioral implication
Type of construction	Wood frame with pitched roof, gypsum board partitions.	Flexible, economical. Homelike appearance, important for normalization (T) "It could be anybody's house. The pitched roof gives the image of shelter." (A)
Floor area	3328 square feet	More storage space, a den or recreation room, basement and attic would have been desirable (H, T).[b]
Exterior finish	Gray clapboard	Color and texture blend in with the surrounding woods. (A)
Plan configuration	U-shaped with main access located inside the "U"; two bedroom wings separated by an outdoor court	This creates three major areas in the house: a "public" area containing living room, dining room, kitchen, laundry and sitting room, and two "private" areas containing the bedrooms and bathrooms. There are no "semipublic" or "semiprivate" areas available, and residents complained that they had no place to go if they wanted to be with one or two others. (R, H) "We didn't want something that was too rambling. We wanted fairly enclosed outdoor space: providing places for people to get together but not be forced together. (A)
Circulation	Runs along the inside of the "U"	Since the sitting room is in the major circulation path, it serves this function. (T)
Relation between components of living area	Spaces are fairly open. Living room is separated from dining room by brick fireplace. Kitchen opens onto dining room, through a doorway and counter. Sitting room is open to living room. Wooden closets separate dining room from entry.	This allows for easy visual and aural contact between spaces. Coordinators can keep an eye on residents, residents can easily see what is going on, but privacy or quiet are hard to come by. (H, T)
Relation between components of sleeping area	Bedrooms are off a single-loaded corridor. Single rooms share a bath, double rooms have their own.	Residents do not have to leave their bedrooms in order to reach a bathroom. (A) This insures privacy. (T)

(In the Plan configuration cell, a diagram of the U-shaped plan is shown with areas labeled "PUBLIC" and "PRIVATE.")

[a] R, residents; H, house coordinators; A, architect; S, administrative staff; B, board of directors; T, research team.

[b] The Architect would have included a basement if the water table had been low enough.

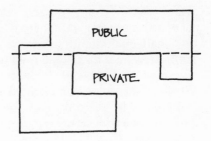

FIGURE 28. Simplified layout of houses
at New England Village.

basement or attic. The "U" shaped plan creates three major areas: a "public" living area and two "private" sleeping areas. A problem is that there are no "semipublic" or "semiprivate" spaces available. Residents complained that there was nowhere to go if they wanted to be with one or two others. Their choice is limited to the "private" space of the bedroom or the "public" space of the living room. In some houses this is a particular problem because the bedrooms are considered off limits to the opposite sex and are often too small to seat a small group. Since the sitting room is in the major circulation path, it does not alleviate this problem. The openness of the living area allows easy visual and aural contact between spaces. Thus, coordinators can keep on eye on residents and residents can easily see what is going on in different areas, but again, privacy and quiet are difficult to obtain.

Use and Evaluation by Space

A narrative description of each area in the house, its use and a capsule evaluation are followed in the report by charts listing design variables, corresponding design decisions, and behavioral implications of those decisions.

House design was also evaluated by three general social–environmental issues: social contact, activity support, and symbolic identification. The house design provided for complete privacy (in bedrooms and bathrooms) and for large-group social interaction (in the living room and sitting room), but was inadequate for pairs and small groups who wanted to interact in isolation.

Analysis of activity support provided by house design focused on six functions: food preparation and consumption, house maintenance, laundry, leisure activities, personal hygiene, and sleeping. In general food preparation and consumption were rated high with adequate room and pleasant facilities. House maintenance was quite easy, although difficult-to-clean materials such as dark tile and light-colored furniture were used in some places, and storage was inadequate. Laundry facilities were adequate. There were some complaints about lack of space and inadequate lighting for arts and crafts activities. The semiprivate bathrooms (one for two people) were well received, yet the rooms were too small for two people at once, making hygiene instruction awkward. Finally, there were complaints that beds were uncomfortable and that noise could be heard through the walls.

The house design fared quite well on "symbolic identification," the second large social–environmental issue. The house was seen as quite homelike. It provided security, comfort and group membership. Many staff and residents, however, felt that the house was more a showplace than a home. The Village received many visitors, and a desire to create a good impression ran counter to other needs to personalize and alter spaces. Evidently most residents and staff see New England Villages as something that is not quite like a home, but is certainly not an institution. The evaluators coined the term "congregate living facility" for this undefined condition.

DISCUSSION

This study is notable from several perspectives. It carefully considered the social–historical context. The evaluators had a good understanding of the history and current philosophy of mental retardation and of the relationship of these trends to the built environment. The evaluators decided that three concepts bridged treatment and built-environment issues: control over one's life, growth, and integration with the mainstream of culture.

The evaluators recognized, however, that these concepts were difficult to measure. Hence, in their evaluation the research team focused on three evaluative issues which were more measurable: social contact, activity support, and symbolic identification. Focusing on these issues permitted a careful space-by-space evaluation which was useful to designers yet which also was closely tied to the Village's operating philosophy.

The study serves as another example of a multimethod approach. Interviews, observations, document reading, unobtrusive measures, and other methods were used. Multiple methods allow cross-checking of information. While each method has its own problems and biases, these weaknesses can be somewhat counteracted by the use of multiple approaches.

In addition, the evaluators were very sensitive to the diverse kinds of people who would be reading their report. The report presents both fine-tuning suggestions for the Village and design suggestions for future projects. There are many graphics directed at designers such as photographs, floor plans, and point-by-point charts. Also, however, there are several sections directed specifically at policymakers.

Ultimately, the excellence of this evaluation stemmed from the evaluators' careful conceptualization of the focal problem (see Table 11 and Figure 27 for examples). Such careful work at the outset of an evaluation can save much confusion at the end.

Charlesview Housing

Study by: John Zeisel and Mary Griffin
Methods used: Observation, interview, analysis of documents
Type of project: Resident evaluation of the designer's intentions at an apartment complex
Information source: *Charlesview Housing: A Diagnostic Evaluation*, 1975

INTRODUCTION

As can be seen by their representation in the present volume, housing or apartment complexes have been prime candidates for evaluation. Charlesview Housing, a 220-unit apartment complex, was evaluated by a research team consisting of members of a graduate design course. This evaluation was particularly notable because both the designers and users had an important role: the users evaluated the success of the designers' expressed ideas.

STUDY DESIGN AND METHODS

The evaluation involved two basic research questions: finding out how design decisions were made and finding out how residents use and think about their homes. The design decisions were analyzed in two steps. First, the evaluators examined the working and final plans, the records of all the parties involved, and their letter files. Second, the architects were exhaustively interviewed about the issues raised in these steps.

The residents' opinions of the project, and their uses of it, were tapped by observing and interviewing them and observing physical cues. Observations of residents were generally qualitative rather than quantitative; they served to suggest relevant issues rather than support or reject hypotheses. A notable exception, however, was the evaluators' recording and statistical analysis of furniture arrangement. Observation of physical cues involved looking for points of conflict or harmony in the environment, as well as for indicators of use or misuse. For example, a "no ball playing" sign indicated to the evaluators that there was a conflict between the area's potential recreational uses and its other demands. The resident interview served as the major data-gathering instrument. It included questions such as "Is there enough privacy in your dining room?" and open-ended questions such as "What do you particularly like about living in Charlesview Housing?" The residents also played a "floor plan game," in which the residents indicated their preferred room arrangements.

These data were collected by a graduate environmental design class during a 3-month period in the spring of 1973.

Setting

Charlesview Housing consists of 212 units of duplex row houses stacked to produce three and four-story buildings. The buildings are constructed out of gray concrete and form structures of different sizes and orientations so as to provide a variety of space and designs.

The buildings are sited on a triangular lot defined by two busy commercial streets. The buildings are inwardly focused. By their placement, they define several small clusters and a large central wall, an attempt to create a hierarchy of spaces, as illustrated in Figure 29.

Physical Context

The inward focus of the buildings is dictated by the commerical nature of the neighborhood, which is dominated by facilities such as car dealerships and warehouses. Although this siting suggests potential conflict between residential and commercial uses, it also ensures that commercial amenities are very accessible.

Social–Historical Context

The present site of Charlesview Housing was originally slated for high-rise apartments. However, since the mid-1960s were a time of considerable turmoil and growing social awareness, contention over the high-rise plans resulted in the appointment of a mayoral "blue ribbon" committee. The committee eventually recommended that housing for moderate and low-income people be built under the federal 221(d)3 program. The sponsor of the project was the Committee for North Harvard, a nonprofit ecumenical corporation, who hired the Pard Team as architects of record.

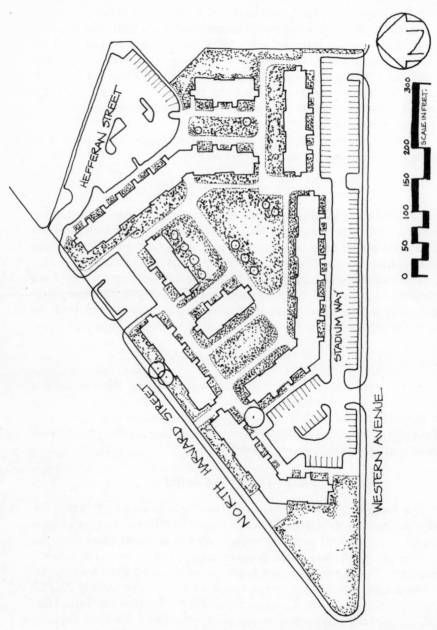

FIGURE 29. Site plan of Charlesview Housing Project.

Since the late 1960s also demanded an increasing fiscal involvement in the Vietnam war, serious cuts were made in the federal housing budget, and concomitant cuts were seen in the Charlesview FHA budget. For example, picture windows on the ground level became much less desirable when privacy fences were cut from the budget.

Design Process

The Pard Team approached the design with an explicitly social orientation; they wished to "create a physical environment which would better the lives of people who would live in the housing." This goal carried the explicit belief in building form as a social influence. One partner stated, "We can create tensions or relieve tensions by what we design."

Except for zoning controls placed on the parcel by the Boston Redevelopment Authority (BRA), Pard Team was not given any special instructions by either the BRA or their nonprofit clients, the Committee for North Harvard. Pard Team prepared alternative designs—from single family to high-rise high density—so that various ideas and issues could be discussed with the Committee for North Harvard. The architects felt that much of their information came from several site visits by their clients.

Users

Charlesview Housing consists of about 35% one-bedroom units, 18% two-bedroom units, 30% three-bedroom units, and 18% four-bedroom units. These apartments helped to dictate the demographic mix of the complex. Over one-third of the residents were singles or couples, and many of these were elderly. There were also many children and large families; over 40% of the apartments had families with at least three children.

Although other demographic information was not available to the evaluators, the interviewers estimated that the complex was about 15% minorities and 85% white and contained mostly middle-class residents.

EVALUATION

The evaluators organized their evaluation of the design decisions in terms of the three overlapping behavioral issues and the several physical dimensions illustrated in Figure 30.

Design Decisions

PHYSICAL DIMENSION \ SOCIAL ISSUE	 Orientation	 Territory	 Privacy
RELATION OF SITE TO SURROUNDINGS	SHARED OPEN SPACES & LOCATION OF BUILDING ENTRIES		
SITE CONSIDERATIONS	LOCATION OF BUILDINGS, OPEN SPACE, & SITE FACILITIES		
RELATION OF SITE TO BUILDINGS	LOCATION OF PARKING AREAS & BUILDING ENTRIES	PLAN OF SHARED OPEN SPACE	
SEMI-PUBLIC SPACES IN & NEAR BUILDINGS		STAIRWELLS & STORAGE AREAS	
SEMI-PRIVATE AREAS OUTDOORS		BALCONIES & PATIOS	BALCONIES & PATIOS
RELATION OF SEMI-PRIVATE AREAS TO DWELLING UNIT			DOORS BETWEEN DWELLING UNITS & PATIOS OR BALCONIES
PRIVATE AREAS: DWELLING UNIT			LAYOUT OF APARTMENTS

FIGURE 30. Design decision matrix.

In general, the orientation resulting from the complex received mixed reviews. Over one-third of the residents rated Charlesview Housing highly for convenience and for ease of directing others to their home (it is directly across from Harvard Stadium). The inward orientation was less positively viewed. The residents didn't perceive the intended "hierarchy of space." The interior walkways weren't frequently used because parking was on the outside of the site and because the inside entrance to the complex's convenience store was blocked off. This latter finding suggested that the architects could have more thoroughly considered the attractions of the parking areas and of the other amenities.

Territory was primarily defined, in order of decreasing scale by: (1) the building siting, (2) the stairwells and storage areas, and (3) by balconies and patios. As was mentioned above, the siting scheme was not well understood by residents, and they didn't use the interior spaces frequently. During the design process, however, much of the landscaping budget was cut, including funds for benches and play equipment. The resulting spaces were much less attractive than those designed. Only six units shared a stairwell, yet the residents didn't feel that they knew the people sharing their stairwell better than they did other people in the complex. The evaluators felt that this was perhaps due to the highly heterogeneous composition of the complex. Finally, the importance of clearly defined yards was striking. Whereas over 80% of families with well-enclosed, small (4' × 8') yards further developed them with planting or furniture, almost none of the units with undefined yard areas chose to develop them.

Privacy also provided mixed reviews for the architect. Sliding patio doors were provided between the living area and the patios or balconies in an attempt to increase the apparent size of the units. This feature was viewed positively by second-floor residents; over 70% said that they used their balconies, and 77% felt that there was enough privacy in their outdoor spaces. On the first floor, however, the privacy fences had been cut late in the design process for reasons of costs. Only 26% first-floor residents used their patios, and only 38% felt that they had adequate privacy. (See Tables 12 and 13.) Many first-floor residents complained that their only choices were either to open their curtains and expose themselves to passersby or to constantly live with closed curtains.

The interior arrangements of the apartments were well received, however. The floor plans provided features such as entry into a foyer, a close relationship between kitchen and eating areas with only a counter separating them, and a more complete separation between the eating/dining area and the living area. The residents' satisfaction with these design decisions were evidenced in the results of the floor plan game, the observation of apartment use and furniture arrangement, and by interviews.

TABLE 12
Use of Outdoor Space

"Do you or your family ever use your outdoor space for sitting out?"	
	Sit outside, %[a]
Balcony	70 (37)
Patio	26 (18)

[a] Number of families on which this percentage is based is given in parentheses.

TABLE 13
Privacy of Outdoor Space

"Do you feel there is enough privacy in your outdoor private space?"	
	Feel privacy is adequate, %[a]
Balcony	77 (37)
Enclosed patio	90 (10)
Open patio	38 (8)

[a] Number of families on which this percentage is based is given in parentheses.

Overall, Charlesview Housing received quite high marks. The interior arrangements provided the greatest sources of satisfaction, and the building siting showed the greatest possibility for improvement. The major complaints, however, such as the lack of privacy in first-floor units and the lack of use of enclosed spaces, were caused at least in part by postdesign cost cutting. Although these could not have been anticipated, the evaluators suggested that architects consider designing-in features which could be cut without ruining the integrity of the entire design. At Charlesview, the possibility remains that these complaints can be remedied by action of the residents or the landlord.

DISCUSSION

This evaluation of Charlesview Housing provides another example of a competent, multimethod project. With limited funding, the research team combined limited direct observation with more extensive interviewing and produced a coherent, useful product.

There was apparently a previously considered research strategy, and this forethought contributed to the final results. The evaluators used an inductive approach, allowing their open-ended examination of documents to suggest further, more detailed questions. Although they used a behavioral process orientation (e.g., examined privacy, territory, and orientation) rather than the five-part heuristic proposed in the present volume, the fact that an organizing heuristic was adopted helped guide the investigation and helped structure the final report.

Within its constraints, this evaluation succeeded well. It was accomplished with limited funding in a few months by a graduate design class. A more thorough evaluation might have allowed, for example, more complete observation of the site, perhaps using one of the behavioral mapping techniques outlined in Chapter V. Also, the one-time evaluation did not allow very careful analysis of the effect of season on living patterns. Finally, the fairly small sample (55 respondents) did not allow for much statistical analysis, perhaps missing trends which an evaluation of a larger sample might have revealed. All of these shortcomings exemplify the sort of trade-offs that occur with a relatively small and inexpensive project.

National Parks Visitor Centers

Study by: Ervin Zube, Joseph Crystal, and James Palmer.
Methods used: Questionnaires with users and staff, interviews with administrators, systematic observations, several checklists, and content analysis of documents.
Type of project: Comparative evaluation of twelve National Park Service visitor centers.
Information source: *Visitor Center Design Evaluation*, 1976

American attitudes toward the outdoors are changing. More and more people are camping, hiking, backpacking and generally spending more time in the outdoors. Although these changing attitudes are affecting many facets of society, from clothing styles to the sales of recreational vehicles, the national parks are bearing the brunt of the changes. Once used only by a few people (who were mostly white and upper class), the parks are now used by a broad range of visitors. Heavy use of the parks has prompted serious discussion as to whether campsites at national parks should be on a reservation-only system, with waiting lists of months or years for a site.

The following evaluation study is a comparative study of twelve National Park Service visitor centers directed by one of the present authors (E. Zube). The study serves to exemplify our approach in several ways (see Chapter I). The study explores the four components of environmental design evaluation: the setting, the proximate environmental context, the users, and the design activity, and views them from a larger social–historical context. The study employs several complementary techniques such as interview, questionnaire, and observation. Finally, the visitor center evaluation is a comparative one, allowing us to see how different centers attack common problems.

INTRODUCTION

Visitor service centers serve as the primary focus of park services in many national parks. They provide historical information in historic-theme parks (e.g., Wright Brothers) and information concerning nature and natural processes in recreational and natural-theme parks (e.g., Yosemite). They provide toilet facilities, sell books and pamphlets, often house park offices, and provide other services. The twelve centers were chosen "so as to provide (1) geographical distribution, (2) representation of natural, historic, and recreation themes, (3) examples from several construction periods . . . and variability in quality . . ." (p. 7). The chosen sites (Figure 31) were Bandalier, New Mexico; Cape Cod Province Lands, Massachusetts; Olympic–Hoh, Washington; Petersburg, Virginia; Rocky Mountain–Alpine, Colorado; Fort Raleigh, North Carolina; Gettysburg–Cyclorama, Pennsylvania; Great Falls Park, Virginia; Rocky Mountain Headquarters, Colorado; Scotts Bluff, Nebraska; Wright Brothers, North Carolina; Yosemite, California. The study was commissioned by the National Park Service (NPS) and was carried out by E. Zube (a landscape architect, planner, geographer), J. Crystal (a graduate student in landscape architecture), and J. Palmer (a graduate student in forestry); a multidisciplinary, university-based team. The study is notable as a model of a comparative evaluation which provides quite rich information with a fairly small time investment at each site—nine person days.

METHODS AND STUDY DESIGN

Budget and time constraints precluded the use of a study design based on random data-gathering procedures. Therefore, days were selected for field studies to coincide as nearly as possible with periods of peak visitation at each center. Peak visitation periods were selected because they represent the conditions under which a majority of all visitors experience the centers. They also represent the time when the facility is most heavily taxed to satisfy visitor needs. Retests were conducted at Wright Brothers and Cape Cod Province Lands in mid-August, approximately 5 to 7 weeks after the initial site visits, to test the influence of time of season on visitor perceptions and patterns of use. These test/retest data suggest that there is little variability between pre- and peak-season ratings at Cape Cod Province Lands related to increases in numbers of visitors. This increase does not apparently influence the perceived quality of the building, exhibits, or audio-visual program. See Table 14 for a summary of data collection methods.

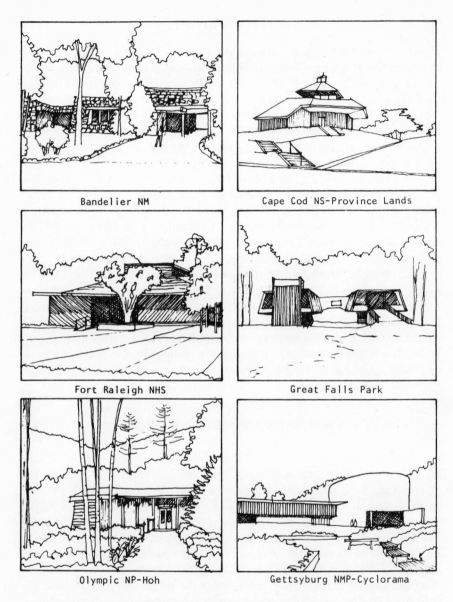

FIGURE 31. Visitor centers at various parks studied (by Joseph Crystal).

Questionnaires were randomly distributed to visitors on the same days on which observation periods were scheduled. The number of questionnaires distributed per hour was, however, generally proportional to the number of visitors per day. The visitor centers were divided

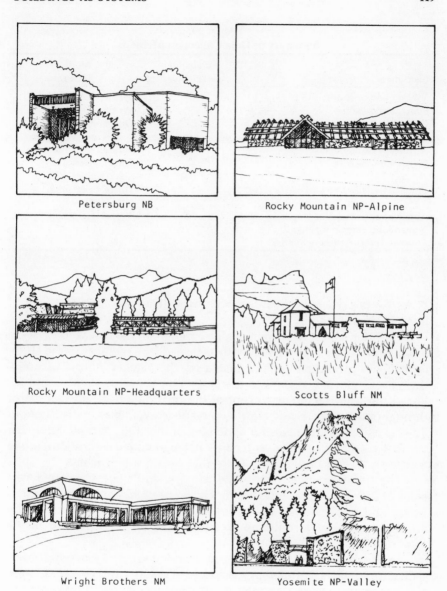

Petersburg NB

Rocky Mountain NP-Alpine

Rocky Mountain NP-Headquarters

Scotts Bluff NM

Wright Brothers NM

Yosemite NP-Valley

FIGURE 31. (continued)

into three groups according to level of visitor use during the peak season, and the minimum number of questionnaires to be collected at each center was determined relative to its visitation level. Staff questionnaires were mailed to pertinent office personnel and were distributed in the

TABLE 14
Summary of Data Collection Methods

Instrument	Number of participants	Mode of gathering
Visitor questionnaires	3065	Written response
Staff questionnaires	150	Written response
Observation code and checklist	257 (observation periods)	Direct observation into coded categories
Architect questionnaires	?	Written response
Barrier-free checklist	—	Direct observation of facilities
Physical facilities checklist	—	Direct observation of facilities
Administrator interview (super-intendent, chief of mainte-nance, chief of interpretations)	36	Verbal interaction

parks by the chiefs of maintenance, park historians, and/or chiefs of interpretation.

Questionnaires were personally handed to visitors as they were leaving the center and subsequently collected by a member of the study team. This procedure greatly increased the rate of return of questionnaires, allowed visitors to ask questions about the study, and allowed the team members to control the distribution of questionnaires throughout the day. Every effort was made to avoid bias in distribution. The normal procedure was for a member of the study team to approach the first visitor leaving the center after having handed a questionnaire to a previous visitor until the quota for that period was reached.

The questionnaires distributed to park staff were shortened versions of the visitor questionnaire to which questions concerning maintenance and opportunities for open-ended responses were added. The questionnaire sent to building architects was open-ended and in the form of a written interview.

In addition to questionnaires and interviews, both systematic observation (normally scheduled at 75-minute intervals), and checklists were also employed in the field analysis. A barrier-free checklist was adopted from Tica and Shaw (1974) and a physical facilities checklist was developed especially for this study. The purpose of such checklists is to guide the study team in making a comprehensive and consistent appraisal at all sites, thus allowing for the comparative analysis of data across sites.

Two limitations on data availability merit mention. Historical data in the form of building plans, master park reports, and related documents were not uniformly available for all parks and centers. Hence, there are gaps in the information as to salient design activity issues for all facilities studied. And finally, no data were available on maintenance and operations costs for the specific buildings and related areas being studied. This is particularly unfortunate as it precludes the inclusion of an important element—cost—in the evaluation of the facilities.

Setting

Data in this section regarding spatial issues were extracted from visitor center plans, systematic observations of visitor use of the facilities, questionnaires, and field notes. The functional distribution of space was generally as follows: approximately 25% of floor area allocated to reception functions, 30% to interpretive exhibits; an auditorium, when present, occupied 20%, leaving 10% for comfort facilities and 15% for office space.

The "intensity of use" was determined for each center. This was expressed in square feet of circulation space per person and was defined in two ways: (1) the daily peaks (the mean from the noon-to-4:00-p.m. period) and (2) the weekly peaks (the second highest observed during the observation period). Perceived crowding, as measured by the visitors' reports on a five-point scale, varied with density, with high perceived crowding (1.27–1.87) occurring at 16–20 square feet per visitor. This is illustrated in Table 15. Perceived comfort, attractiveness, and "invitingness" did not appear to vary with density, however, as is illustrated in section The Users below.

The number and types of alterations to the centers were taken as an index of the fit of the original buildings with their functions and activities. These changes have often resulted from increased use of the parks. Additions included increased exhibit space at Scotts Bluff and Bandalier. Alterations included expansion of the sales areas at many of the centers.

Maintenance issues were explored by an interview with the chief of maintenance or his counterpart at each center. Although the centers were exceptionally well maintained, maintenance problems were exacerbated by the location of parks in extreme climates. Four general categories of problems were identified: (1) soft ceiling materials, (2) leaking and snow removal problems on flat-roofed buildings, (3) heat loss due to noninsulated glass or single entry doors, and (4) inadequate heating and ventilation systems. Other problems included the use of nonstandard fixtures and building materials which were difficult to replace, large lawns to mow, and floor surfaces that were difficult to clean.

TABLE 15
Perceived Crowdedness[a]

Park[b]	Overall	Daily peak use
Bandelier NM	1.68	1.63
Cape Cod NS–Province Lands	1.83	1.82
Fort Raleigh NHS	1.36	1.24
Gettysburg NMP–Cyclorama	1.27	1.29
Great Falls Park	1.28	1.27
Olympic NP–Hoh	2.52	2.70
Petersburg NB	1.36	1.44
Rocky Mountain NP–Alpine	3.13	3.34
Rocky Mountain NP–Headquarters	1.72	1.87
Scotts Bluff NM	1.45	1.39
Wright Brothers NM	2.19	2.30
Yosemite NP–Valley	1.87	1.91
Average	1.86	2.10

[a] Range in value on a 5-point scale (1 = uncrowded to 5 = crowded).

[b] NM, National Monument; NP, National Park; NHS, National Historic Site; NMP, National Military Park.

A barrier-free checklist was adopted to assess the accessibility of the centers for persons with handicaps. This checklist focused on several areas such as the parking area, curb cuts, walks, ramps, restrooms, etc. A listing of the ranking of all centers is presented in Figure 32. Finally, there was considerable concern with the safety and security of the centers. Although few accidents occurred, superintendents and maintenance personnel reported several concerns: security of footing, vehicular accidents, and site-specific problems such as rattlesnake bites at Great Falls and lightning at Rocky Mountain–Alpine. Security is an increasing problem. Break-ins were reported at several centers, resulting in the purchase of alarm systems and an increase in security systems at some centers.

The Context

Four issues were primarily relevant to the context of the centers: "(1) the park theme, (2) the extreme environmental conditions, (3) spatial characteristics, and (4) regionally oriented aesthetic perception of users."

The centers generally have two themes, depending on the park, either natural (e.g., Rocky Mountain–Alpine) or historical (e.g., Wright Brothers). Although all centers were seen as suitable, the interiors and exteriors of historic-theme centers were perceived as being more suitable than the natural-theme centers. This is perhaps because the historic-theme centers did a slightly better job of relating their themes to the building and exhibit design.

VISITOR CENTER	PARKING AREA	CURB CUTS	WALKS	RAMPS	ENTRANCE	DOORWAYS	STAIRS	FLOORS	REST ROOMS	DRINKING FOUNTAINS	PHONES	RANK ORDER
BANDELIER NM	L	L	P	P	L	P	P	P	P	P	NA	11
CAPE COD NS-PROVINCE LANDS	P	P	A	P	A	A	P	P	P	P	L	1
FORT RALEIGH NHS	P	L	A	P	L	L	L	A	P	L	NA	11
GETTYSBURG NMP-CYCLORAMA	L	L	A	P	A	A	P	P	L	P	L	4
GREAT FALLS PARK	P	P	A	P	P	P	P	A	P	P	L	1
OLYMPIC NP-HOH	L	L	P	P	P	A	NA	A	P	P	L	6
PETERSBURG NB	P	A	A	P	L	A	P	A	P	P	L	6
ROCKY MOUNTAIN NP-ALPINE	L	L	P	L	A	P	NA	P	P	P	NA	1
ROCKY MOUNTAIN NP-HEADQUARTERS	P	L	A	L	L	A	P	P	L	A	L	9
SCOTTS BLUFF NM	P	A	P	L	A	A	A	P	P	P	L	9
WRIGHT BROTHERS NM	L	L	A	P	P	P	P	A	P	P	L	6
YOSEMITE NP-VALLEY	NA	L	P	P	A	A	P	A	P	P	L	4

KEY: A=ADEQUATE L=LACKING P=PARTIALLY ADEQUATE NA=NOT APPLICABLE

FIGURE 32. Visitor center barrier-free access.

The nature of national parks dictated that centers be placed in extreme or fragile environments (e.g., Olympic receives over 200 inches of rain per year; Rocky Mountain–Alpine is sited in the tundra). These conditions required special roof and structural considerations in the centers. The fragility of the environments necessitated special attention be given to circulation patterns and crowd-handling modifications such as a fence at Rocky Mountain–Alpine.

Three types of placement were seen for the visitor centers: near the park entrance, en route between the entrance and the possible destination, and at a terminus. Each location had its own challenges and requirements. For example, a center at the entrance had to set the mood for the park, while a terminal center had to provide a synopsis of park values.

Other park influences were important as well, requiring consideration of the larger context. For example, the entrance road to Olympic passed by a forest clear-cut area. An opportunity for education was apparently lost here because the public was not educated about the purposes of clear-cutting.

In order to judge aesthetic preference and to understand regional building styles visitors were asked to indicate which building materials they thought most appropriate for a new building if the one they were in was destroyed. Visitors generally: (1) chose materials similar to those present for replacement, (2) chose rough-textured over smooth materials (i.e., rough-hewn wood over smooth-finished wood) when given a choice.

The Users

A demographic profile of the users was obtained by questionnaire. Most were young (mean age 31.9) and well-educated (63.4% had at least some college education). Most respondents were either not in the labor force (41.6%) or were in managerial or professional roles (36.3%). Also, most (83%) had visited at least two national parks in the previous 3 years. The majority (70.2%) of visitors arrived at the parks by car, although this ranged from a high of 95% at the Rocky Mountain centers to a low of 63.8% at Fort Raleigh. Interviews revealed that, contrary to expectations, the reason people went to the centers was not principally to use the restrooms. Rather it was to look around (36.4%), obtain information (24.3%), and view exhibits (21.7%). Only 4.3% listed use of the restrooms as their primary purpose for visiting the centers.

Visitor and staff perceptions of several areas of the building (e.g., interior, exterior, exhibits, services, arrival area) were solicited by asking respondents to judge the areas on adjective pairs such as colorful–colorless, suitable–unsuitable, or ordered–chaotic. Visitor ratings are illustrated in Figures 33 and 34 as examples. Both staff and visitors were able to qualitatively differentiate between the different areas. It is interesting to note that ratings by staff were consistently lower than those by visitors. This is perhaps due to their continued exposure to the centers.

The Design Activity

While this study focused primarily on design functions, it reviewed the planning function as well. The planning and design process was followed quite closely in four of the twelve centers.

VISITOR CENTER	ARRIVAL AREA			BUILDING EXTERIOR		BUILDING INTERIOR					
	CONVENIENT	SAFE	ATTRACTIVE	ATTRACTIVE	INVITING	COMFORT	ATTRACTIVE	ORGANIZED	LIGHT	INVITING	QUIET
BANDELIER NM	1.41	1.51	1.41	1.31	1.49	1.50	1.51	1.48	2.04	1.60	1.64
CAPE COD NS-PROVINCE LANDS	1.28	1.28	1.26	1.25	1.30	1.51	1.28	1.36	1.32	1.44	1.96
FORT RALEIGH NHS	1.27	1.19	1.15	1.23	1.33	1.20	1.26	1.33	1.42	1.30	1.50
GETTYSBURG NMP-CYCLORAMA	1.29	1.25	1.29	1.50	1.59	1.42	1.42	1.36	1.50	1.50	1.50
GREAT FALLS PARK	1.46	1.41	1.73	1.35	1.41	1.71	1.51	1.61	1.89	1.61	1.41
OLYMPIC NP-HOH	1.46	1.54	1.51	1.65	1.63	1.90	1.62	1.52	1.80	1.67	2.39
PETERSBURG NB	1.20	1.18	1.20	1.19	1.32	1.20	1.25	1.27	1.33	1.25	1.41
ROCKY MOUNTAIN NP-ALPINE	1.50	1.69	1.63	1.47	1.58	1.89	1.78	1.65	1.87	1.87	3.07
ROCKY MOUNTAIN NP-HEADQUARTERS	1.33	1.40	1.42	1.40	1.50	1.45	1.48	1.52	1.68	1.60	1.80
SCOTTS BLUFF NM	1.29	1.26	1.44	1.43	1.51	1.37	1.52	1.41	1.51	1.48	1.53
WRIGHT BROTHERS NM	1.29	1.33	1.41	1.29	1.59	1.48	1.42	1.30	1.40	1.51	1.77
YOSEMITE NP-VALLEY	1.86	1.54	1.61	1.53	1.90	1.78	1.62	1.54	1.99	1.74	2.03
RANGE (HIGH-LOW) ON A 5-POINT SCALE	.66	.59	.58	.46	.34	.69	.53	.38	.72	.62	1.66

FIGURE 33. Perceived quality of arrival areas and buildings by visitors. Entries represent the means from five point scales. 1 is most positive and 5 most negative.

VISITOR CENTER	EXHIBITS						AUDIO-VISUAL					
	INFORMATIVE	COLORFUL	INTERESTING	ATTRACTIVE	STIMULATING	ORGANIZED	INFORMATIVE	COLORFUL	INTERESTING	ATTRACTIVE	STIMULATING	ORGANIZED
BANDELIER NM	1.91	1.54	1.46	1.37	2.02	1.56	1.22	1.47	1.52	1.43	1.61	1.46
CAPE COD NS-PROVINCE LANDS	1.54	1.85	1.49	1.69	1.96	1.56	1.57	1.48	1.56	1.60	1.98	1.70
FORT RALEIGH NHS	1.54	1.78	1.50	1.50	1.96	1.57	1.48	1.91	1.69	1.99	2.12	1.62
GETTYSBURG NMP-CYCLORAMA	1.34	1.47	1.32	1.47	1.72	1.42	1.42	1.45	1.40	1.50	1.76	1.46
GREAT FALLS PARK	2.36	2.41	2.15	2.12	2.82	2.10	1.60	1.59	1.57	1.71	1.93	1.51
OLYMPIC NP-HOH	1.39	1.70	1.43	1.58	1.89	1.50	NOT APPLICABLE					
PETERSBURG NB	1.22	1.99	1.17	1.59	1.55	1.35	1.19	1.39	1.14	1.36	1.48	1.28
ROCKY MOUNTAIN NP-ALPINE	1.58	1.60	1.44	1.55	1.88	1.50	NOT APPLICABLE					
ROCKY MOUNTAIN NP-HEADQUARTERS	1.62	1.83	1.68	1.81	2.25	1.74	1.40	1.50	1.45	1.49	1.74	1.55
SCOTTS BLUFF NM	1.55	1.65	1.36	1.53	1.72	1.36	NOT APPLICABLE					
WRIGHT BROTHERS NM	1.25	1.91	1.33	1.57	1.88	1.40	1.30	1.67	1.36	1.55	1.60	1.53
YOSEMITE NP-VALLEY	1.44	1.81	1.55	1.68	2.03	1.59	1.05	1.84	1.81	1.83	2.26	1.63
RANGE (HIGH-LOW) ON A 5 PT. SCALE	1.12	.94	.98	.75	1.27	.75	.46	.52	.07	.61	.78	.42

FIGURE 34. Perceived quality of exhibits and audiovisual programs by visitors. Entries represent the means from five point scales. 1 is most positive and 5 most negative.

Site selection was accomplished either by implementing the master plan (four sites), by the building designers (three sites), or by other personnel (five sites). A comparison revealed that sites established by the master-planning teams were better at integrating the various needs of the designer and chief of interpretation.

Five centers were designed through the efforts of a multidisciplinary design team rather than a single consultant designer. The qualitative advantage of the multidisciplinary approach was seen in improved relationship between building and site such as the providing of visual orientation cues or curvilinear parking area forms echoing the lines of the centers.

Finally, although no conclusive data were available, it appeared that better exhibits resulted when exhibit designers were included in the process from the beginning.

DISCUSSION

Directed by one of the present authors (E. Zube), this study, as might be expected, fits our proposed model for environmental design evaluation quite well. In the introduction to this study we pointed out some notable aspects of the study; they deserve some further elaboration here.

(1) Multiple measures were used which contributed to the success of the study in some important ways. Qualitative, open-ended interviews helped the evaluators define the problem and helped to suggest issues that might not have been raised otherwise. These qualitative methods supplemented the more concrete quantitative observational techniques. Also, whereas each method had its own bias and inherent problems (e.g., verbal responses might be biased in interviews or the presence of an observer may alter a behavior), overlapping methods were used such that the problems with one technique might be compensated by the strengths of another method.

(2) This study compared several centers, allowing the relative strengths and weaknesses of the various centers to be seen in contrast. For example, if the Rocky Mountain–Alpine center were the only center evaluated and visitors rated it an average of 3.13 on a 1 to 5 scale of "crowdedness," what conclusion can be drawn? When we examine ratings of the other sites, however, we see that most hover between 1.25–1.9, and we can conclude that Rocky Mountain–Alpine is quite crowded.

(3) Each aspect of our four-part environmental design model (e.g., setting, proximate environmental context, users, design process) was carefully addressed. This approach provided a structure for defining the problem and for assuring that the various important influences (some of which, such as maintenance personnel, are frequently not considered) *were* adequately sampled. By helping to define the problem, the scheme also helped to establish the study design and to suggest methods. For example, a focus on the design process dictated that the relevant architects should be interviewed, and that the design program should be studied.

With this study, as with all studies, certain cautions are in order. One caution relates to the self-selection of the visitors. When we compare questionnaire responses between the centers, it is not entirely clear whether the differing responses are due to the centers themselves, or because different types of people pick different centers and those people respond differently. For example, souvenir sales were rated much lower in Olympic-Hoh than in other parks. Was this a deficiency in Olympic's souvenir sales, or did souvenir haters simply go to Olympic? We can gain some insights by looking at our other data (Olympic visitors were the most highly educated of all parks) and by trying to elicit a comparative appraisal from each visitor (i.e., we can ask how this center compares with other centers the visitor has seen). The present study used this comparative appraisal in some instruments and not in others.

Caution should be used with the variability of the data. For clarity, visitor and staff responses were analyzed in terms of percentages. This sort of analysis is useful but we must be cautious in its application. For example, if one respondent rates souvenirs a "5" and another respondent "1," the mean of these ratings is "3." We should have less confidence in the mean as a reasonable summary of the ratings than if each respondent rated the souvenirs a "3," also resulting in a mean of "3." It is important to describe the data in statistics which are based on variability of the data.

IV

Outdoor Spaces

This chapter focuses on the evaluation of outdoor places and spaces. The five studies discussed in the following pages include an urban plaza, an urban park, a portion of a college campus, a planned unit development (PUD), and multiple housing sites. The first three studies are evaluations of outdoor spaces while the latter two are more appropriately described as evaluations of building-site complexes. While a case can be made for their inclusion in Chapter III, major emphasis in these two studies is on the quality of the site planning and design.

The differences among the kinds of settings discussed in Chapters II, III, and IV sometimes seem small and almost insignificant. The subjects of the three chapters do, however, parallel the usual division of design responsibilities among the professions of interior design, architecture, and landscape architecture. There are also several other differences which tend to differentiate the environments described in these three chapters, including scale, boundary definition, temporal use patterns, and user identification.

The scale or areal extent of most interior spaces is expressed in square feet. Individual buildings are also described in terms of square feet and frequently in terms of height or number of stories as well. While the scale of some outdoor spaces, such as miniparks, may be expressed in square feet, the more common areal unit is the acre or hectare (2.47 acres). The scale of the projects in this chapter ranges from approximately one acre in the urban plaza to 125 acres in the PUD. As the project being evaluated increases in size, the evaluator can be limited with respect to the use of some techniques such as observation, where logistical problems can become overwhelming if the entire site is to be considered. Given a large area, the evaluator may be confronted with the need to sample the environment as well as the users. In other words, random spatial-sampling techniques or sampling of specific functional areas can be employed to make the problems of large areas more manageable and to accommodate the use of techniques such as observation. The housing-sites study, for example, employed observation but limited its use to children's play areas.

Boundary definition can also be a more complex problem for many outdoor space studies than for interiors or individual buildings. While the actual geographic extent of a park or plaza, and the delineation of its edges, may be quite explicit, the space is also more vulnerable to natural and man-made external or physical contextual factors. There are no walls to limit the lateral transmission of sound from roadways as, for example, at the PUD and some of the housing sites, or ceilings to intercept the falling of rain, sleet, or snow.

Lateral boundaries can also be highly variable with regard to ease or difficulty of visual and physical penetration. The urban plaza and park present somewhat contrasting conditions. The plaza is generally open visually and physically, providing ease of access, while the park is perceived by some as secluded and closed.

Outdoor spaces frequently have more variable patterns of use than do buildings. Some parks or plazas may have entirely different patterns of use from season to season or during the course of a single day. Environmental conditions are much less susceptible to control and therefore more deterministic with regard to uses and activities. Amount of sunlight, temperature, wind, and precipitation all influence use of the out-of-doors. Thus the evaluator must not only be particularly attentive to these and other external or contextual influences, but must also be sensitive to the temporal problem and the need for longitudinal studies to get a truly representative evaluation of outdoor places and spaces.

The users of apartments, dormitories, institutions, schools, and offices are relatively easily identified. They are the individuals who live, work, or study there. They can usually be identified by name (and address where necessary) from the normal kinds of records that are maintained at such places. Users of many outdoor spaces are difficult to identify. Categorizing users by observation provides one means but can present problems of reliability. More important, however, is the identification of nonusers and the inquiring as to their perceptions of the place and reasons for nonuse. Both the urban park and the college campus studies attempted to address this problem. The urban park researchers polled nonusers (with questionnaires) in surrounding office buildings, and the college campus study placed a "mail-back questionnaire" in the campus newspaper. While both contributed limited, but important and otherwise unavailable data, the direct approach in the park study was the more successful.

The settings for the five studies are predominantly urban or suburban. The college campus site, which is urban in character and use, is in a rural context. The PUD and one of the housing sites are also more suburban than urban. All of the multifamily housing sites are located in England.

The study designs range from the pilot or exploratory nature of the urban plaza study to the well-defined instruments and sampling procedures of the housing sites, and from the case-study approach of the former to the multiple-site, comparative approach of the latter. The purposes of the studies include: (1) the development of data for design programming (college campus site), (2) assessment of designers' assumptions (urban plaza and PUD), (3) residents' satisfaction with alternative forms of multifamily housing (housing sites study), and (4) a combination of the assessment of design assumption and data for redesign (urban park study).

An important note about the design and conduct of evaluation studies has been provided by the author of the urban plaza study (Rutledge, 1975). Rutledge suggests that important "design-relevant" insights evolve out of informal debriefings of researchers immediately after their return from the field. Such debriefings help orient the analysis of data, to suggest particular issues and areas for closer scrutiny, and also to flag issues and areas which might otherwise be overlooked because of the absence of "a lot of" data. Rutledge suggests that there is much to be learned from anecdotal information acquired through sensitive eyes and ears—information not always obtainable by more structured social science research methods. The studies in this chapter deal with such anecdotal information in varying degrees, with the urban park and plaza studies perhaps being most successful. The park researchers used such anecdotal information initially as a means to design and develop more structured methods.

How have the data from these several studies been used? Have they provided feedback to the design process? The answer to these questions, as one might expect, is mixed. The most successful, however, were those that were undertaken with a particular designer/client in mind, the PUD and the college campus site. The PUD study changed some ideas and procedures in the professional office that sponsored it. The data from the evaluation of the campus site fed directly into the design program for the area.

Evaluation of a Campus Space

Study by: Howard Cohen, Joseph Crystal, Jessie Pflager, Richard Rosenthal, and Hollis Wheeler

Methods used: Observation, newspaper questionnaire

Type of project: Design program data

Information source: *Design Evaluation of a Central Outdoor Space at the University of Massachusetts,* 1976, and interview with Nicholas Dines, Landscape Architect, Amherst, Massachusetts

INTRODUCTION

This study is perhaps more appropriately described as an evaluation for design. It is an assessment of a major outdoor space on the University of Massachusetts campus, a space that had been transformed over a period of approximately 10 years without benefit of a design to guide its evolution. The objective therefore was to assess users' perceptions and patterns of use for inclusion in the program for the future design of the space. An important element of consideration in the assessment was the long-standing planning assumption as to the physical and social significance of this central campus space adjacent to the Student Union.

The study was undertaken by five graduate students at the University, one each from sociology and psychology and three from landscape architecture.

STUDY DESIGN AND METHODS

Data were collected by questionnaire (Figure 35 shows an excerpt) and systematic observation. The questionnaire was designed to be printed in the campus student-run newspaper, *The Collegian*. It appeared in the paper on a midweek day, Tuesday, with the intention of reaching the widest possible range of the University's population. The paper has a circulation of approximately 20,000.

How would you rate the space using the following adjective pairs?

```
                 1      2      3      4      5
uninviting    ( )    ( )    ( )    ( )    ( )  inviting
comfortable   ( )    ( )    ( )    ( )    ( )  uncomfortable
ordered       ( )    ( )    ( )    ( )    ( )  chaotic
alien         ( )    ( )    ( )    ( )    ( )  friendly
suitable      ( )    ( )    ( )    ( )    ( )  unsuitable
ugly          ( )    ( )    ( )    ( )    ( )  beautiful
```

During an average week in the spring, how often do you engage in the following activities in this space?

ACTIVITY   (Number of Times)

a)   arrange to meet someone there   _____
b)   eat lunch there   _____
c)   ride a bike there   _____
d)   walk through it   _____
e)   run or jog through it   _____
f)   stop there to talk with someone   _____
g)   study or read there   _____
h)   receive a notice or a pamphlet there   _____
i)   show the space to an out of town friend
       as being a symbol of the campus   _____
j)   go there to "hang out"   _____
k)   ride a motorcycle through it   _____
l)   play frisbee or other games there   _____
m)   listen to or observe entertainers   _____
n)   purchase something from a crafts vendor   _____
o)   park my car there   _____
p)   other (please explain)   _____

_____

_____

_____

Of the above activities, please list the five you would most like to do in the space. (You may include any others that you think are appropriate).

FIGURE 35. Sample of questionnaire printed in student newspaper.

The questionnaire included four primary parts: a set of six semantic differential scales, an activity checklist, a format for self-reports on primary and secondary circulation pathways, and a checklist for determining the respondent's relationship to the University. Questionnaires were collected by means of collection boxes in the lobbies of the Student Union and Campus Center, both buildings adjacent to the study site. A total of 104 completed questionnaires were received.

Systematic observation employed the use of a Super 8 (8-mm) movie camera located on the seventh floor of the University Library

FIGURE 36A. Area of campus space study in foreground (courtesy of the University of Massachusetts Photographic Service).

which bounded the site on the south. A time-lapse technique was adopted using color film and exposing one frame every two-thirds of a second or approximately 91 frames per minute. Five-minute film sequences were taken during the midpoint of seven 15-minute class-

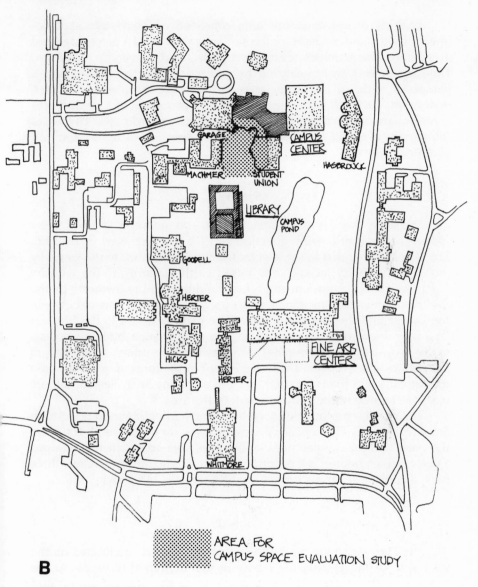

AREA FOR
CAMPUS SPACE EVALUATION STUDY

B

FIGURE 36B. Site plans showing relationships of area of study to the rest of the campus.

change periods between 9:30 a.m. and 4:00 p.m. and during the midpoint of seven 50-minute class periods starting at 8:55 a.m. Filming was conducted on two midweek days in April, a Wednesday and a Thursday.

Analysis of questionnaire data consisted of descriptive statistics (mean, mode, and median) for the semantic scales and a percentage and distribution for activities, circulation paths, and university affiliation. Time-lapse films were analyzed by projecting selected frames into a perspective grid of the study area and locating the position of each pedestrian. Each cell in the perspective grid was equal to a ground area of ten square feet on a side and was identified by coordinates. The use of the perspective grid facilitated analysis of direction of movement and pedestrian density.

Setting

The study space is rectangular in form and measures approximately 250 feet in the north–south direction and 175 feet east–west (Figure 36). The space is bounded to the east by the Student Union, to the south by the 27-story library located on a raised earth platform, to the west by Machmer Hall (a classroom and office building), and to the north by the Campus Center parking garage and a portion of the open space between the garage and the Campus Center. The main entrance to the Student Union is oriented to this space. In addition to such building access, major pedestrian-way entrances are at each of the corners. About 60% of the surface area is covered with pavement in the form of sidewalks and the remains of a roadway that once traversed the area. Several mature trees are located in the southerly half of the area.

The space is undoubtedly one of the most intensely used pedestrian areas on campus. It is a major pedestrian way for north–south traffic on the west side of the campus. It also provides access to the Campus Center, which houses, for example, four separate dining facilities, three bars, a book store, and a host of student services and activity offices.

Context

The Student Union and the adjacent study space are located on the west side of the geographic center of the University campus. Major features of this central space, in addition to the Student Union and library on the west, include the Fine Arts building on the south, a roadway on the east, and the Campus Center on the north. The midpoint is occupied by the campus pond and surrounding lawn areas.

Dormitories are located generally to the northeast, southeast, and southwest. The southwest complex, housing some 5000 students, is a major contribution to the north–south traffic pattern through the space.

EVALUATION

Of the 104 respondents to the questionnaire, approximately 78% were undergraduates, 16% graduate students, 3% faculty and staff, and 3% other. Statistics describing responses to the semantic scales are given in Table 16. Obviously these users perceived the space as predominantly uninviting, uncomfortable, chaotic, unsuitable, and ugly. The one scale that indicated a more moderate response, but still not positive was alien–friendly.

Analysis of the responses to the question about activities engaged in in the space indicate that very few of the respondents used it as anything more than a place to walk through. Over 82% never met someone or played games there, over 93% did not use it as a place to eat out-of-doors and 99% did not use it as a place to "hang out." However, 68% reported they talked with someone there one or more times per week, presumably while walking through it to a class or dormitory.

TABLE 16
Semantic Scale Responses

Scale	Mean	Mode	Median
Uninviting–inviting	2.01	1.0	1.94
Comfortable–uncomfortable	3.67	4.0	3.77
Ordered–chaotic	3.77	5.0	4.07
Alien–friendly	2.52	3.0	2.50
Suitable–unsuitable	3.70	5.0	3.95
Ugly–beautiful	1.93	1.0	1.78

TABLE 17
Usage Preferences

Potential use	First choice (%, $N = 82$)	Second choice (%, $N = 80$)
Meet someone there	23.8	3.7
Eat there	21.4	17.5
Ride a bike there	8.3	2.5
Run or jog there	0	6.3
Walk through it	27.4	22.5
Talk with someone	1.2	20.0
Study or read there	3.6	5.0
Show space to a friend	1.2	3.7
Hang out there	3.6	5.0
Play a game there	1.2	5.0
Listen to entertainment	4.8	2.5
Purchase an item	0	2.5
Park a car there	1.2	1.2
Other	2.4	2.5

First- and second-choice responses to the question about what respondents would like to do there are indicated in Table 17. At least for these individuals, it would appear that the space could provide a wider range of alternatives and be multipurpose in nature. The social activities of meeting a friend and engaging in a discussion as well as having a place to eat out-of-doors and just walking through were most frequently seen as desirable uses.

Figure 37 illustrates the pedestrian path analysis from the ques-

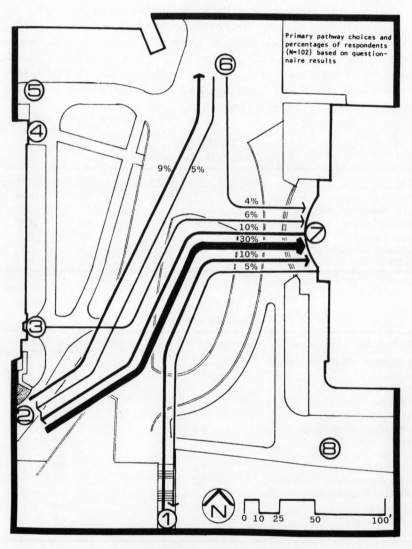

FIGURE 37. Pedestrian path analysis.

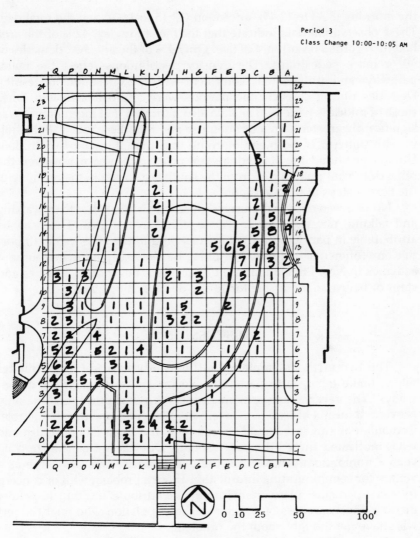

FIGURE 38. Map showing pedestrian-density grid.

tionnaire data. Percentage figures indicate use of primary pedestrian paths. Line thicknesses also indicate intensity of use. The Student Union accounts for 65% of the combined primary origins and destinations and 60% of the secondary.

The major outdoor access point to the space is at the southwest between the library and Machmer Hall which accounts for 53% of the primary path and 39% of the secondary path entries and departures.

Pedestrian density maps were prepared for each film time period for each grid cell (Figure 38). Summary density maps were also compiled for the morning (8:30 to 12:45), afternoon (12:45 to 4:00), and the total day. These observation data indicate that for the study day, 42% of the area had no recorded occurrence of use. Only 4% of the area had densities of 30 or more pedestrians. The density distributions reflect the major pathways and are generally consistent with the questionnaire results. Densities during class-change periods were approximately ten times those of midclass periods. The data also indicated that morning use was significantly greater than afternoon use and that there was a slight shift in path pattern. In both the morning and the afternoon the Student Union constituted one of the dominant poles in the space; however, the other dominant pole shifted from the southeast corner in the morning to the library steps in the afternoon.

No one was observed engaged in any activities other than walking and talking, nor was bicycle use observed in the area. This may be attributable in part to both the time of year and the absence of supportive environmental features. April may be a bit early for many outdoor activities in New England, and the space does not contain benches and steps or balustrades suitable for sitting on or leaning against.

DISCUSSION

The two methods employed in this study have several attributes which make them particularly suitable for an "evaluation for design study." The newspaper questionnaire serves a dual purpose: it not only provides a means of user response, but also serves an educational role. Certainly one can and should speculate on the small rate of return and what motivated those who did respond. One cannot generalize from such a nonrandom sample. But the questionnaire can also serve as a vehicle for communicating information about a problem with or concern for some particular place or environmental attribute and thus hopefully increase the awareness of a sector of the population who might otherwise have not thought about the problem.

A method such as the unobtrusive observation used in this study is particularly important to provide a cross-check on nonrandom questionnaire data. The use of movie films as an observational means of recording behavioral data has the added advantage of providing a record which can be returned to again and again for additional or more detailed analysis.

One obvious limitation of this study is the short time period during which the data were collected. One might profitably speculate on patterns of use during other seasons of the year and at night. But such cross-sections of time, even though limited, do provide a kind of information in a form which would not otherwise be available to the designer charged with responsibility for the transformation of places.

Design Activity

The data collected in this study have been used by the landscape architect, Nicholas Dines, in the development of his design for the space in both direct and indirect ways. In an early study in which the pathway and density analysis data have been spatially delineated together with existing physical features such as the trees, data were found to be most valuable for the designers in specifying the nature and magnitude of a functional problem to be resolved. It enabled the designer to get to the stage of plan refinement and the studying of details sooner and with a better understanding of the problem.

The landscape architect reported that the study also prompted him to think about design strategies and techniques that he may not have discovered without the study. In other words, it was a stimulus to the creative process of giving form to the environment. The pulsating nature of the pedestrian traffic flow and, in particular, the short but high-intensity bursts between class periods suggested the need for visual accessibility into the space from the more constricted entry points at the southwest and the southeast, entry points from which visual access was partially blocked by large, low-branching trees and variations in surface elevation. Two strategies which he considered to help alleviate this problem were, first, to raise the vegetation canopy from an existing height of approximately 7 or 8 feet to approximately 15 feet and, second, to simplify and clarify ground plane elevations and to allow for paving patterns to provide the sense of direction and flow. These are rather direct design strategies, but their potential utility was reinforced by consideration of the behavioral data.

Furthermore, the information of desired multiple-uses from the questionnaire provided guidance for the design of peripheral areas. Overall, the data helped to remove some of the inevitable uncertainty about preferences and behavior and enabled the designer to focus more rapidly on physical form.

The designer also was aware of the shortcomings of the data and would have liked (1) more observation over varying seasons; (2) additional analysis of the relationship of sun and shade and morning and afternoon time periods to spatially shifting circulation patterns; and (3) detailed behavioral analysis of eddying patterns which might occur as pedestrians move from more constricted linear entry paths into larger, less well-defined circulation spaces.

First National Bank Plaza

Study by: Albert J. Rutledge
Methods used: Observation, interview, review of documents
Type of project: Assessment of designers' assumptions and design objectives
Information source: *First National Bank Plaza, Chicago, Illinois, A Pilot Study in Post Construction Evaluation*, 1975, and correspondence with A.J. Rutledge.

INTRODUCTION

A conventional approach to environmental design evaluation has been that of the professional critic, who assesses a particular setting, drawing upon insights gleaned through perceptual processes particularly attuned to designed environments. This critic is perhaps aided by professional design training, with verbal communications skills and probably an occasional bias (to which we are all equally susceptible). The assessment, more often than not, focuses on the formal, physical, visual attributes of the setting. It is usually addressed to professionals and published in one of the professional design journals, or occasionally appears in the Sunday supplement of major newspapers in a more popularized version.

This study involved an urban plaza in Chicago (Figure 39) that had been the subject of a number of such conventional evaluations. The plaza, adjoining the First National Bank, had been acclaimed "a successful design" by critics. Positive comments focused on materials and forms and suggested that it had been "built for people." It was hoped that the user-oriented study reported here would add to the understanding of the success or failure of the plaza by looking more closely at the user dimension. In addition, such an evaluation could also possibly shed some light on the validity and generalizability of more conventional approaches to evaluation.

Objectives of the study, while not explicitly stated, were to (1) test the designers' assumptions about users and user patterns of behavior, (2) assess user response to the plaza, and (3) compare the findings

142

FIGURE 39. First National Bank Plaza (photograph courtesy Arthur Kaha). The First National Bank Plaza in Chicago, Illinois was designed by the Perkins and Will Partnership, Architects, and Novak and Carlson, Landscape Architects.

from a user evaluation with those from a conventional "critic" type evaluation.

The designers' assumptions about the three-level plaza were the following (Rutledge, 1975):

The overall space was designed to:

1. draw a variety of people . . . no particular "type" . . . just anyone and everyone.
2. serve basically as a through walk or access way to peripheral buildings and/or a place to sit, watch and relax.
3. be flexible . . . service its basic functions as well as be adaptable for special events as exhibits and entertainment programs.

The intermediate level was meant as:

4. a place for observing the fountain and happenings on the lower level.

The lower level was planned so as to:

5. minimally encumber spontaneous behaviors. (p. 9)

The study was conducted by Albert J. Rutledge, associate professor of landscape architecture and five graduate students, Mark Brenchley,

Robert Callecod, Louis Messina, Stephen Qualkinbush of Landscape Architecture and Arthur Kaha of Architecture at the University of Illinois, Urbana-Champaign.

STUDY DESIGN AND METHODS

This was intended to be a pilot study. The researchers, therefore, warned that their effort should be viewed "less as an evaluation of the plaza than as a display of what was done and how it was accomplished" (p. 2). The nature of a pilot study is to investigate methods, refine data-gathering instruments, and point the way for future studies. Nevertheless, the findings suggest that the pilot study tapped salient issues and provided valuable evaluative information, as well as addressing procedural and methodological issues.

Methods employed in this study included document review, observation, interviews with designers and users, and photographic documentation. Document review involved an analysis of the plans for the plaza in order to abstract some of the obvious design intentions and to provide antecedent data for interviews with the designers. Interviews were essentially open-ended but guided by the questions listed in

TABLE 18
Questions for Designers

1. Who specifically was the client on this project? With whom did you deal during the design development stage? How would you describe your relationship?
2. Were any other professionals involved in the design?
3. What were the design criteria for the project?
4. Who established those criteria?
5. Who decided finally what would be included in the project?
6. Why were these things included?
7. Were there any special considerations or restraints imposed which were uniquely important to the final design? (funds, city codes, etc.)
8. Who did you expect to use the project? Why?
9. Would you please describe how you expected people to use the various parts of the site? How was this established?
10. What special design considerations were necessary to satisfy the intended users and their activities?
11. Have you been to the site since it has been completed? Were people using it as you had anticipated?
12. How do you feel about the site as built? How do you think the users feel about the site?
13. What parts do you think function best? Worst?
14. Are there things happening on the site which you had not anticipated, and feel should or would have been designed for?
15. If you had the opportunity would you change any parts of the site? Why?
16. Would you change your design process if you were commissioned to do another project like this one?

TABLE 19
Steps for Observations

1. Refer to station-selection chart for point of beginning.
2. Follow subject (selected from random-selection chart) through plaza until exit; i.e., when puts foot down on first step or when door opens.
3. On the map, record time of entry at entry point and time of exit at exit point.
4. In space provided, record personal data for each observation (subject).
5. On the map, graphically record movement and activity of each subject. See symbol legend.
6. Time each major behavior and record adjacent to symbol on map.
7. If, at a given station, the subject as indicated on the random-selection chart does not appear within three minutes then follow the next subject. If no subject appears with a five minute span, then circle the station, record a "no show" and proceed to the next station.
8. If "next subject" as noted in No. 7 above is similar to the last subject observed, then pick the next dissimilar subject and follow through the plaza.
9. Sign each map in appropriate space.
10. When changing "guard" (each hour) the new observer will follow the station-selection chart, continuing from the last station of the previous observer.
11. Behavior symbol for *other* ($\bigcirc$) should, when used, have a number in the circle and the activity referenced and noted in the margin of the map.

Table 18. Plan analysis and interviews served to identify the sampling of designers' assumptions previously noted and which were judged to be sufficient in number for a pilot study.

Observation provided a systematic, albeit indirect means of assessing user responses to the plaza. Observation data addressed the questions of: Who is there? What are users doing? Where are they? and When are they there? Two observation strategies were employed. The first involved the taking of 35-mm color slides at 15-minute intervals (two or three exposures each time) from the fifth floor of an adjacent building. The slides were to be used to determine densities of people in the plaza. The second strategy involved the tracking of randomly selected users and the mapping of their locations. Table 19 illustrates the steps involved in this procedure. The goal of this strategy is to identify representative patterns of users by aggregating the patterns of randomly selected individuals (see Figure 40).

User interviews ($N = 80$) were intended to augment observation data, were brief, and consisted of the following open-ended questions:

1. Why do you come to this plaza?
2. What do you usually do here?
3. Where did you come from to get to the plaza?
4. What do you like most—and least—about the plaza?
5. How would you modify this plaza to make it better?

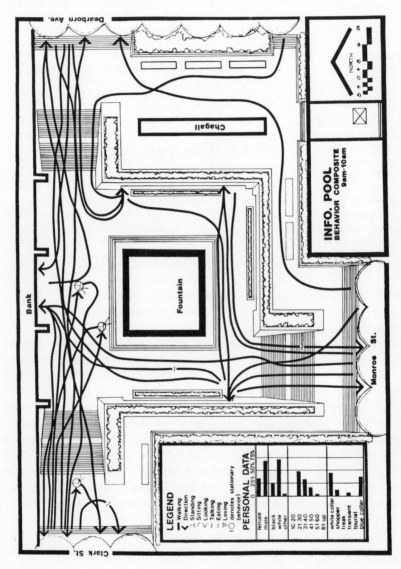

FIGURE 40. Primary paths through the plaza (courtesy Arthur Kaha).

In addition, the personal characteristics of the people interviewed were noted, as were the characteristics of the people usually tracked, namely age, sex, race, and type of person (e.g., executive, white-collar, shopper, blue-collar, "freak" or "hippie," and tourist).

The original research design called for a sampling of "typical days and times in the life of the plaza." This procedure was not followed, however, due to a lengthy period of inclement weather. Instead, three visits were made to the site on separate days with each visit encompassing one-third of a "study day" running from 8 a.m. to 5 p.m. When combined, the data from the visits approximated one full day.

Another change made in the design as a result of "real world" experience affected the user interviews. Interviews which were originally the responsibility of the observers were reassigned to employees of the adjacent bank because of time demands on the observers.

A final change in the study design involved expanding the role of the photographers to include: (1) taking a panoramic series of slides every hour to assist in identifying the most popular areas; and (2) recording specific confrontations or interactions of users with the environment. It was hoped that these additional data would compensate, at least in part, for the abbreviated sampling of days and times.

Setting

The plaza encompasses three levels and is approximately one acre in extent. The first level coincides with the surrounding streets and consists of tree-lined sidewalks on three sides which provide access to the broad flights of steps leading to levels two and three at lower elevations. The second, or intermediate, level contains a number of levels and a large Marc Chagall mosaic in the eastern section (see Figure 39). The lowest level is characterized by a large central fountain. This level also provides access to a bank on the north of the plaza, a coffee shop in the west, and a shopping–subway concourse on the east. This level is surrounded on all four sides by window walls which permit viewing from the adjacent interior areas.

Context

The plaza is located in southcentral Chicago in the "loop," the city's main business district. It is bounded on three sides by narrow traffic-congested streets which separate it on the east and west from prestigious office buildings and on the south from lesser structures. Within the city block on which the plaza is located is also found a fifty-seven-story bank building to the north, a lower building on the west, and a one-story disguised cooling structure on the south which also serves as the base for three large flagpoles. An underground parking garage is on the east of the plaza beneath the intermediate level.

Users

Who used the plazas during the time data were collected? Users were predominantly white males characterized as white-collar workers. In addition, the majority of the users who completed the questionnaire administered by bank employees were observed to be under the age of 30. The distribution of users during the day is indicated in Figure 41. Not surprisingly, the lunch hour accounted for the vast majority of plaza users.

EVALUATION

Given the pilot nature of the study, evaluation is considered first from the vantage point of the plaza and second from the vantage point of the study design. Figures 42–44 summarize the findings from the questionnaires. For example, dislikes included the temperature in the plaza, entertainment programs (both frequency and kind), crowding, the outdoor cafe, and inadequate seating. The attributes that were most liked included the entertainment programs, the fountain, people watching, and the atmosphere, isolation, and appearance of the plaza. The desired changes in the plaza mentioned most frequently reflected these likes and dislikes. They included providing more seats, more programs, and more greenery. The latter presumably might also provide shade from the hot sun. Reasons for going to the plaza included entertainment, for people watching, relaxing, using the bank, and eating lunch. Intentions and actual behavior appear to be highly correlated as indicated by the responses in Figure 44 to the question, What do you usually do here?—watching, listening, eating, sitting, and relaxing.

Were the designers' assumptions about users and patterns of use realized? The sample of users who responded to questionnaires suggests that the plaza does not draw a variety of people and in fact attracts a particular type—young, white, male, white-collar workers. This finding could, however, be a result of biased sampling, of a predominantly male social context in the surrounding business environment, or of a design which does not support the needs of diverse groups (e.g., the many steps might discourage older users).

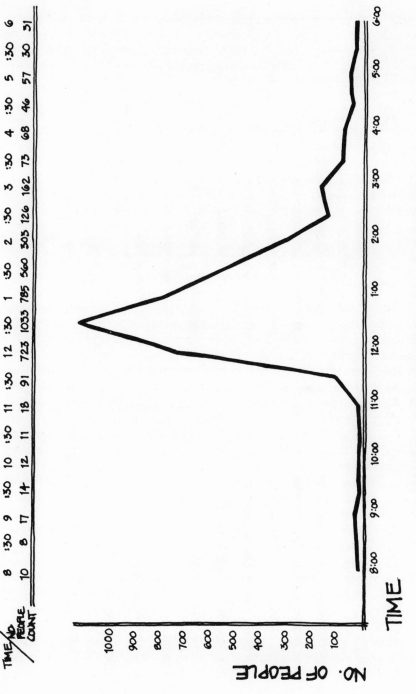

TIME	8	:30	9	:30	10	:30	11	:30	12	:30	1	:30	2	:30	3	:30	4	:30	5	:30	6
No. PEOPLE COUNT	10	8	17	14	12	11	18	91	723	1053	785	560	505	126	162	75	68	46	57	30	51

FIGURE 41. Distribution of people using the plaza according to time of day.

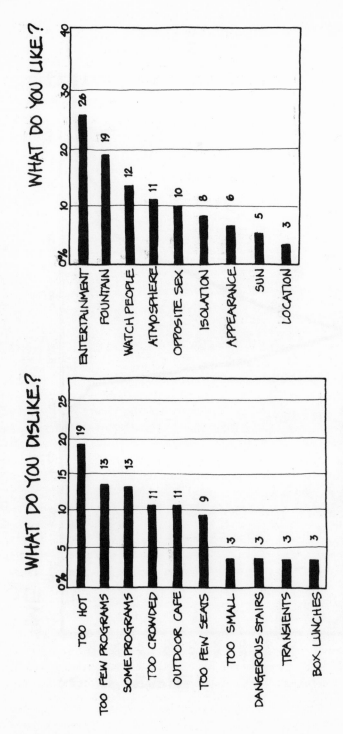

FIGURE 42. Results of questionnaires: what do you like? dislike?

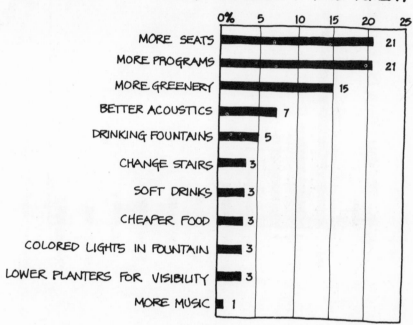

HOW WOULD YOU CHANGE OR MODIFY THIS PLAZA?

FIGURE 43. Results of questionnaires: what would you change?

The tracking and observation data indicate that the plaza does serve as both a place to walk through and is a place to sit, watch, and relax. As a through route, one of the most frequented paths was down one flight of steps, past the fountain, and up on another side of the plaza. The edges of the steps and the planters were also heavily used for sitting.

The adaptability and flexibility of the plaza for special events as well as for basic functions of circulation and relaxation was found to be limited. It should be noted, however, that observation of special events (e.g., entertainment programs and a temporary outdoor cafe in the vicinity of the Chagall mosaic) was informal and not systematic. These activities were seen to hinder normal traffic flow. The overall site design was defined as essentially "a number of aisles" with no "subspaces" to accommodate such activities.

The intermediate level of the plaza was judged to function as intended for observing the fountain and people. However, design details such as the height of plantings and the locations of other users tended to limit or block sight lines. Little data were obtained which provided for an evaluation of the lower level as a place for spontaneous behavior.

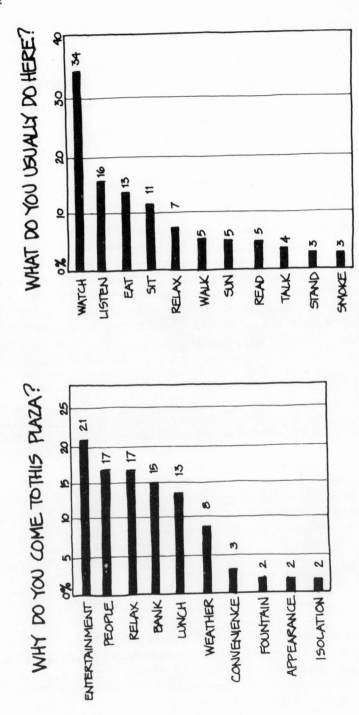

FIGURE 44. Results of questionnaire: why do you come here? what do you do?

A retrospective examination of the study led the researchers to a number of conceptual, procedural, and methodological conclusions. (1) The study did indeed provide factual information that were not available via a traditional assessment by a design critic and which is in a quantitative form that provides for easy comparison of findings from different studies. (2) In the very early stages of the study design, an iterative process of defining objectives, identifying methods, and pretesting helped to refine procedures and reveal questions and problems which might otherwise have complicated and compromised the conduct of the study. An important adjunct of this consideration was that the use of multiple methods provided for a certain amount of redundancy, allowing one method to check another—a necessary test of validity. (3) Problems in the current study emphasized that the focus of a study should be specified and limited and thus realistic of accomplishment. Further, the focus should derive from the design objectives and the designers' intuitions and/or assumptions. (4) Time sampling for observation should include a range of days, conditions, and times. The use of predetermined coding categories for observation provides a specific example of the need for pretesting, as the observers in this study had a difficult time discriminating between executives and white-collar workers. In addition, the collection of observation data by multiple means (e.g., tracking, photography, and area observation) could provide more detailed data if they were coordinated in time of collection so as to allow for comparisons among data sets.

Finally, several conclusions emerged with regard to the use of interviews. A two-stage procedure should be employed. This would provide for (1) open-ended questions with a very general response which could increase understanding of the setting and lead to more detailed questions, and (2) focused questions which address the specific evaluation issues which form the focal problem. Also, had the interviews followed initial observations, a better sampling strategy could be devised. For example, subjects could be selected so that the sample has a representative number of each kind of user. Furthermore, interviews should include other users of the space such as maintenance personnel, employees of adjacent shops, and program administrators. Interviews also should include nonusers.

DISCUSSION

An important and obvious attribute of this study is its avowed pilot or exploratory orientation. As indicated in the researcher's own evaluation of the procedures and methods, the data are limited. It would in fact be dangerous to draw specific conclusions on the basis of such a limited data base—limited particularly in terms of time sampling. Whether the kinds of users, intensities, and patterns of behavior recorded would prevail in other seasons, on weekends as well as week days, under varying climatic conditions, or in the evening as well as the day are important but unanswered questions.

Another important point of consideration of observation data which was not addressed explicitly in this study is interobserver reliability. For example, what would the correlation be between data sets from two or more simultaneous but independent observations?

The study demonstrates the importance of identifying design objectives and design assumptions as one of the starting points for the evaluation process. This is certainly one of the most important steps in the development of a feedback system of evaluation data for future design decisions. The fact that all of the participants in the research were designers explains, at least in part, this concern for and sensitivity to the utility of findings.

Finally, this study begins to illustrate that user evaluation and traditional design-critic assessments can be reinforcing. However, they provide different kinds of data and insights. Had more information been available to the research team, a detailed analysis of some of the reviews by design critics might have provided the basis for a more systematic comparison between the approaches to evaluation. Nevertheless, the study is unusual in this endeavor.

This study is an important contribution because of its dual focus—on a designed place and on the nature of the evaluation activity itself.

Urban Park Evaluation

Study by: Anita R. Nager and Wally R. Wentworth
Methods used: Questionnaire, interviews, behavioral mapping, participant observation, archival search.
Type of project: User and nonuser evaluation of urban open space.
Information source: *Bryant Park: A Comprehensive Evaluation of Its Image and Use With Implications for Urban Open Space Design*, 1976

INTRODUCTION

The primary objective of this study, undertaken at the request of Community Planning Boards of New York City, was to develop recommendations for improving the design of Bryant Park (Figure 45). Originally designed in 1934, the park offered a unique opportunity to assess the relevance of design principles and assumptions from the 1930s in the context of user and nonuser perceptions in the 1970s.

The study was conducted by two environmental psychologists from the Center for Environment and Behavior Studies at the City University of New York.

STUDY DESIGN AND METHODS

The study employed the following methods: behavior mapping, interviews within the park, written questionnaires in offices adjacent to the park, document review, and participant observation. All data were collected in 1974 during the season of park use, June 15 to August 31.

Behavioral mapping, to find out who was doing what at which location, provided place-specific information. The park was divided into 21 subareas, and users were systematically recorded by age, sex, activity engaged in, location, and physical park feature being used. Thirty-six

FIGURE 45. Bryant Park, New York (courtesy of W. Wentworth and A. Nager).

observation periods were scheduled at 1½-hour intervals to obtain data for all times of the day and all days of the week.

Interviews were conducted with 213 users in the park. The number of users sampled was proportional to the number of people using the park at different hours of the day as determined by the behavioral mapping data. This resulted in a sample heavily weighted with lunchtime users. The interviewer entered the park at a different entrance for each interview round. He or she selected every fifth stationary park user until the time-period quota was met. Interview questions were a combination of fixed and free responses and rating scales. Primary interview topics included use of open space generally and Bryant Park specifically, origin and destination of interviewee, reasons for coming to the park, thoughts about changes and improvements in the park, perceptions of safety, frequency of use, place of residence, and occupation.

The written questionnaire was distributed to employees of several businesses in the vicinity of the park. Of those who responded, 147 were park users and 47 were nonusers. The questionnaire was similar in content to the interview. Respondents, however, represent an opportunity sample (i.e., an easily available group) and not a random sample, thus limiting the generalizability of the data.

The review of historic archives, newspaper files and the like provided information on the uses and design history of the park and, most importantly, a means of ferreting out the objectives and goals of the 1934 design.

Participant observation provided qualitative data based on informal observations and conversations with park users. It was also valuable for the design of the more structured components of the study (questionnaire, behavior mapping codes).

The quantitative data from behavior mapping, interviews, and the questionnaires were analyzed primarily on the basis of the frequencies of various responses. Frequency distributions were tested for significance (i.e., the effect of chance was analyzed) using the Chi square statistic.

Setting

Located adjacent to the New York Public Library in midtown Manhattan, Bryant Park occupies about 4.6 acres of the 9-acre site which includes the library. The park is surrounded by a wrought iron fence and granite wall which matches the granite used in the library. The wall and fence effectively limit ground-level visual access and control physical access into the park from the surrounding streets. The Sixth Avenue

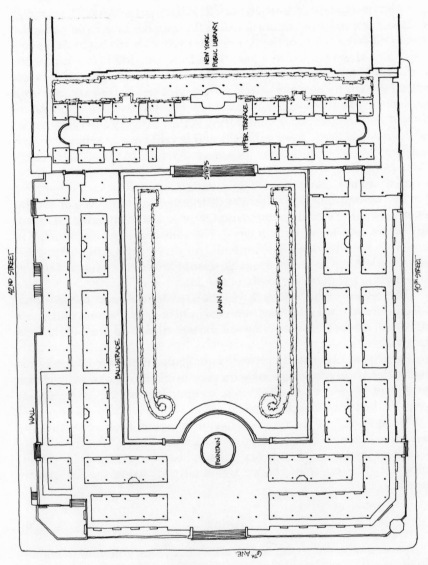

FIGURE 46. Bryant Park: site plan.

entrance on the west side of the park is the only one which permits a view of much of the park and is axially oriented to the library facade.

As seen in the site plan (Figure 46), the center of the park consists of a large lawn area. To the east is the library, separated from the main body of the park by a raised tree-lined terrace with benches. To the north and south are found a geometric arrangement of tree-lined walkways which subdivide these areas into smaller sitting places. Similar sitting places are located in the southeast and southwest corners, while the main entrance with a fountain on axis are located in the center of the west side.

Context

The park is surrounded by high-rise office buildings. Its midtown Manhattan location is devoid of similar open spaces, except for the smaller spaces, primarily along Sixth Avenue, that function primarily as paved plaza settings for major office buildings.

Design Activity

Document review of archives made possible a general reconstruction of the design process and the underlying assumptions for the plan which was executed in 1934. The park had existed since 1884. Its renovation was in response to concern that it had become a poorly maintained plot of ground that was an eyesore in the heart of the city. The condition of the park was a topic discussed in a number of New York papers of the time.

The major assumptions which conditioned the design of the park appear to be:

1. The park should be a setting of "restful beauty," a place for quiet relaxation, a green sanctuary.
2. The park should not provide shortcuts for pedestrian traffic.
3. The park should create a pleasing and complementary setting for the library.
4. Sitting spaces within the park should be designed so as to facilitate a sense of privacy.

In addition to these design assumptions, another major factor entering into the design was the severely depressed state of the économy during the 1930s and the need to keep costs down. The wall around the park was constructed from materials on the site; the fountain had been a part of the earlier design and was simply relocated.

The only major physical change since 1934 has been the modification of the northwest entrance to the park to accommodate a memorial and statue. Other smaller or nonphysical changes included increased lighting, increased police surveillance, and the programming of events in the park such as lunch-hour concerts.

Users

Data on the users of Bryant Park come primarily from systematic observation. Table 20 indicates the proportion of park users of different ages for both weekdays and weekends. Figure 47 shows the distribution of users by time of day. The difference in weekday and weekend use is an obvious reflection of the commercial and business character of the surrounding area, a factor which also contributes to the peak around the lunch hour. This pattern of use varies, however, for the elderly, who remain in the park for a longer period of time in the afternoon.

Figure 47 indicates also that a large portion of the total users are young or middle-aged males. Nager and Wentworth, the researchers, suggest that the lower use by females, as well as the elderly, may reflect perceptions of a lack of security in the park. This point will be addressed in the following section.

The majority of the users in the interview sample, 59%, came from within a three-block radius of the park. Of the total sample, 81.6% traveled six blocks or less. Only 4.1% traveled fourteen or more blocks.

The activities in which the users engaged are indicated in Table 21. Talking, sitting, reading, and eating, all of which are rather quiet and passive park uses, are by far the most popular weekday activities.

TABLE 20
Proportion of Park Users of Different Ages

Estimated age	Weekday, $n = 6411$		Weekend, $n = 2713$	
	f	%	f	%
1–17	171	2.7	58	3.1
18–35	2743	42.8	966	35.6
36–59	1331	20.8	623	23.0
60+	686	10.7	665	24.5
Unknown[a]	1480	23.1	401	14.8

[a] Age breakdowns were not obtained for observations of crowds on the central lawn area and on the upper terrace immediately surrounding the concert events. In these areas simple head counts were taken. The error introduced by this simplification has the primary effect of underestimating attendance in the 18–59 age range but probably has little effect on percentages for age 60 and over.

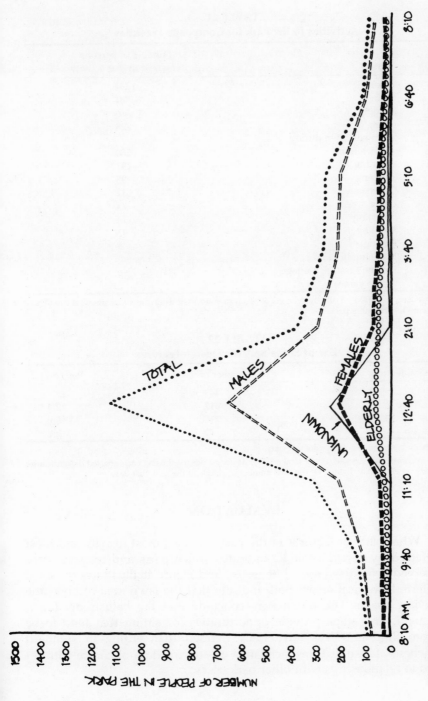

FIGURE 47. Number of persons in the park on a composite weekday by sex and age group. This excludes the lawn area. Because of use of the lawn at certain hours, accurate backgrounds were not obtained for that area. For this reason, the number of "unknowns" is high at the lunch hour.

TABLE 21
Activities in the Park for Composite Weekday[a]

Activity	Number of persons engaged in that activity
Talking	1034
Sitting	584
Reading	366
Eating	226
Sitting on lawn during lunch hour	200
Standing	131
Sleeping	95
Quiet games	31
Lying down	25
Soliciting	14
Vending	13
Buying	9
Listening to radio	7
Picking through garbage	6
Active games	2

[a]These data are based on all observations of behavior conducted during a composite weekday. These data are for the entire park.

TABLE 22
Use of Some Specific Design Features[a]

Physical feature	Number of users	Percentage[b]
Lawn	449	15
Balustrade	412	12.4
Steps	218	6.6
Fountain	72	2.2

[a]Data are based on one composite weekday.

[b]The percentages are based on the total number of people observed using each selected design feature during an entire day divided by the total number of people in the park that day.

EVALUATION

What parts or features of the park are used most heavily and most preferred by users? Table 22 indicates which parts and features were observed being used most intensively and Figure 48 illustrates the most preferred physical areas. Both indicate that the lawn area receives the heaviest usage. The researchers conclude that the balustrade, lawn, fountain, and steps provide opportunities for sitting that tend to be "more fluid and less territorially defined than sitting in standard benches." These features facilitate a greater freedom of choice with regard to relationships with other park users.

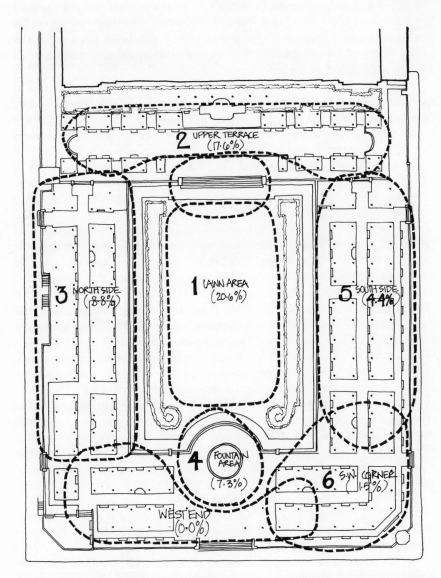

FIGURE 48. Bryant Park: use of physical areas compared. Percentages given of people preferring specific park areas (some expressed no preference, and hence the total does not represent 100% of the subsample of 68 park users).

A subsample of 68 park users was asked if they had favorite areas in the park and if so why and where they were. The reasons given for preferring certain areas included convenience, sanctuary from people or noise, safety, comfort, proximity to music, and physical attributes such as view, greenery, or water.

Content analysis of interview and questionnaire responses to the question, "What three words would you use to describe the park?" shed light on differences in user ($N = 181$) and nonuser ($N = 44$) images of the park. Nearly one-half (42.5%) of the users but only 2.3% of the nonusers saw it as a peaceful, quiet, relaxing retreat. Considerable numbers of both users (34.6%) and nonusers (25.0%) perceived natural qualities of the park to be green, cool, open space. For others, users and nonusers almost equally, the park is a social setting. Users in general were more positive about the park and nonusers more negative.

Perceived safety of the park varied significantly between users and nonusers, with 79% of the users and only 15% of the nonusers rating it safe or quite safe. About one-fifth (19%) of the users saw the park as unsafe or very unsafe while 63% of the nonusers shared that perception. There was no significant difference between male and female park users; however, significantly more female than male nonusers perceived the park as unsafe. The reasons stated most frequently by female nonusers for avoiding the park were "undesirable characters" and "unsafe" (44.4% and 40.7% respectively). These same reasons were given by park users for avoiding certain areas within the park; however, the frequency of response was much lower (12.5% and 6.2% respectively).

Both users and nonusers were asked open-ended questions about changes that would improve the park. About one-fifth of those answering were apparently satisfied with the status quo. Users and nonusers alike desired (in approximate descending order) better maintenance, the planting of flowers in the park, more comfortable benches, restrooms, drinking fountains, and game or picnic tables. The percentage of users and nonusers, respectively, suggesting these changes ranged from 26.1 and 25.0% for maintenance to 4.3 and 5.0% for tables. Approximately twice as many nonusers (91%) as users (41.6%) desired the exclusion of "undesirables."

Both users and nonusers were also asked whether they favored, were opposed to, or neutral to a list of hypothetical changes in the park. All changes were endorsed by a majority of respondents ranging from a low of 52.3% of the users and 53.6% of the nonusers favoring picnic tables to a high of 96.7% of the users and 100.0% of the nonusers favoring a police call-box. The next most favored change by both groups was the addition of chess tables. Nonusers were more inclined to be neutral.

DISCUSSION

Two aspects of this study warrant particular attention. First is the employment of multiple methods of data collection and second is the attempt to include nonusers as well as users of the park in the study.

Multiple methods have provided a richness of data and depth of understanding as well as built-in checks on reliability and validity of findings that would not have been possible with a single-method study design. Perhaps one of the most obvious benefits is the ability to relate what people actually do as determined through systematic observations with their perceptions and attitudes elicited in responses to interviews and written questionnaires. Furthermore, observation data alone would not have provided insights into the different perceptions of users and nonusers nor would it have provided guidance as to the kinds of changes desired for the park.

Understanding of designers' assumptions is an important ingredient in evaluation studies, particularly when the study is of a single site or facility. While multiple sites allow for comparative evaluations among sites—in other words, the evaluation scales can be relative to similar environments being evaluated—single site evaluations obviously do not, and the designers' assumptions and intentions provide a touchstone from which to derive evaluative criteria. Without the archival search, the design assumptions for Bryant Park would not have been known, and an understanding of why the park is the way it is would have been missing from the study.

The inclusion of nonusers should be attempted in evaluation studies of public places. As demonstrated in the Bryant Park case, however, it is far more difficult to deal with nonusers than users. Who are and where are the nonusers are sticky questions that compound problems of study design and in particular of sampling. Nevertheless, the inclusion of even the non-random-opportunity sample in this study provided valuable information.

Housing Site Evaluation

Study by: Ingrid Reynolds and Charles Nicholson, The Housing Development Directorate, Department of the Environment, England.

Methods used: Systematic observation, interview, physical measures

Type of project: Comparative evaluation of six housing exteriors and site developments

Information source: *The Estate Outside the Dwelling*, 1972

INTRODUCTION

The Estate Outside the Dwelling is a comparative study of six housing projects located in London and Sheffield, England. The major objective of the study was to learn how residents from family units in different stages in the family life cycle reacted to different building forms and aspects of site design. The focus of the study was, as suggested by the title, on the exterior of the buildings and the spaces between them. While not explicitly stated in the report, another important implicit objective was to provide guidance to local housing authorities as to characteristics of housing design that enhance resident satisfaction.

The study was conducted by the Housing Development Directorate of the Department of the Environment, which consists of architects, sociologists, other building specialists, and administrators. It is concerned with promoting higher standards of housing through the study of residents or users and of the building process.

STUDY DESIGN AND METHODS

The housing projects and building forms were selected by a team of architects and sociologists. An attempt was made through the selection of projects to have specific building forms replicated at least twice. The

Do you hear noise from traffic at all?   Yes ___
                                                         No ___
                                                  Don't know ___

If yes, where does it come from: (Open ended)

Would you say that getting sunlight in your living room is. . .

                                         Very important ___
                                       Fairly important ___
                                        Doesn't matter ___
                                           Don't know ___

Is your balcony (garden) used for children's play?

                                                   Yes ___
                                                    No ___
                                            Don't know ___

If not, why don't you use your balcony (garden) for
children's play?   (Open ended)

Would you like to have a private garden?   Yes ___
                                                  No ___
                                          Don't know ___

If yes, why would (if no, why wouldn't) you like to have
a garden?
                                          (Open ended)

Summing up your feelings about your house/flat/maisonette
would you say that you are . . .

                                        Very satisfied ___
                                      Fairly satisfied ___
                                 No feelings either way ___
                                    Rather dissatisfied ___
                                      Very dissatisfied ___
                                            Don't know ___

Would you say that the play provision on or near the estate
for children under 11 was . . .   Satisfactory ___
                                              All right ___
                                          Unsatisfactory ___
                                            Don't know ___

Do you ever feel lonely?   Yes ___
                                                  No ___

Do you feel lonely often or only occasionally?
                                                Often ___
                                    Only occasionally ___
                                           Don't know ___

Why do you think you feel lonely?   (Open ended)

FIGURE 49. Sample question from housewives' interview.

forms were also selected as typical examples of their kind. The range of projects and replication of forms satisfied requirements for a comparative study design.

The primary method employed was a structured interview of a random sample of 1317 housewives and 369 of their husbands. The interview data were supplemented with data from systematic observation of children's play in every part of the projects' sites. The housewives' interview included 101 questions, both forced-response and open-ended (see Figure 49 for sample questions). The major topics covered in the interview are indicated in Table 23. Interviews with the husbands lasted about a quarter of an hour and were concerned with car parking issues. Personal background information was also gathered on each of the interviewees. This information was used to define household categories for purposes of data analysis.

Physical measures were also taken of certain environmental attributes such as balcony and garden sizes, heights of fences, daylight in the residences, and distances to car parking areas. These data provided a means for relating specific perceptual and attitudinal responses to design details.

Interview data were analyzed by two methods—content analysis of open-ended questions, and two statistical techniques, correlation and

TABLE 23
Primary Interview Topics

Family relationship (wife, husband)	Likes and dislikes about exterior of projects
Family size—ages	Outdoor family activities
Occupation	Patterns of children's play by age group
Previous housing experiences	Attitudes toward size of play space, location, safety
Reason for moving to present residence	Children's spare time behavior
Perceptions of noises	Car ownership
Views from residence	Accommodations for visitors
Privacy	Social contacts with other residents
Sunlight in residence	Feelings of loneliness or nervousness
Use of balcony or garden	Attitudes toward and use of supporting facilities (i.e., shops, pubs)
If no balcony or garden, desire for same	Perception of project density
Doing of laundry and drying	Perception of project attractiveness
Refuse disposal	Maintenance
Storage	Vandalism
Heating system	Desire to move
Preference for living on ground or on upper stories	Car-parking behavior and attitudes toward parking facilities
Attitudes toward balcony/deck access/elevator and/or stairs	

regression analysis. Correlation analysis indicated the strength of association between pairs of items or variables from the interview, and regression analysis indicated the relative strength or association of multiple items (independent variables) to users' satisfaction (dependent variable) with the "estate outside of the dwelling."

Setting

Table 24 indicates some of the major physical characteristics of the six projects including density, parking, building form, outdoor space, and support facilities. The context of the surrounding environment is also indicated. Density ranges from a low of 51 persons per acre (126 per hectare) to a high of 200 persons per acre (494 per hectare). The provision of parking spaces ranges from .2 to .91 spaces per dwelling units. Building forms vary considerably and include 1- to 3-story row or terrace houses, rectangular 14- to 22-story slab blocks with internal corridor access, rectangular 3- to 14-story blocks with external or outdoor (balcony and deck) access and square point blocks of 11 and 21 stories.

The provision of private outdoor spaces varies considerably from all row-house units with private gardens (Fleury Road) and all apartments with private balconies (Park Hill) to projects in which most units are without private outdoor space (Winstanley Road). Variability in the provision and treatment of public outdoor space is equally as great.

Finally, some of the projects include a range of support facilities within the housing development while others have none. The most frequent facilities are shops, pubs, and laundry drying areas.

Context

The six housing projects vary considerably with regard to the larger environments in which they are located (Figures 50–55). The Fleury Road Project, located on the outskirts of Sheffield, was built on a steeply sloping site overlooking a large wooded open space in a semirural area. The nearest shopping area is about 1½ miles away. Sceaux Gardens, Acorn Place, and Winstanley Road are all in the London Metropolitan area and located in areas that are predominantly residential, while Canada Estate in London and Park Hill in Sheffield are located in mixed residential, commercial, and industrial areas. The Winstanley Road Project is also bounded for a distance of about 250 meters by railroad tracks and a railroad station while Canada Estate is adjacent to a supervised and equipped play field.

TABLE 24
Context and Setting

Project location, Date of construction	Context	Density[a]	Parking[b]	Building form[c]				Setting		Support facilities
								Outdoor space		
				Row houses	Slab blocks	Point blocks	Balcony access blocks	Private	Public	
Fleury Road, Sheffield 1962	Semi rural	51	.91	148/2				All units	Small play area	
Sceaux Gardens, London 1960	Urban residential	136	.20	/1	2/15		4/6	Some balconies	Central open space	
Acorn Place, London 1963	Urban residential	136	.28	/2 & 3			17	Row houses with gardens, most apts. w/o balc.	Extensive paved courtyards	Pub, shops, laundries
Winstanley Road, London 1966	Urban residential	154	.68		1/22	3/11	3/4 & 5	Some gardens and balconies, mostly w/o	8 play areas, much paved space	Shops, children's library workrms, drying facility
Canada Estate, London 1964	Urban mixed	161	.30			2/21	5/3 & 4	3- & 4-story blocks only w garden/balc.	3 play structures	Drying facilities
Park Hill, Sheffield 1961	Urban mixed	200	.18				0[d]/4-14	All units have balconies	5 play areas, large grass areas	Pubs, laundry, police-substation

[a]Density is given as bedspaces (or persons) per acre.
[b]Parking is indicated as percent of dwellings serviced by parking spaces.
[c]The left-hand number indicates the number of individual structures and the right-hand number indicates the height in stories. When no left-hand number is indicated, the structure is usually continuous in form and not in discrete units.
[d]Continuously linked deck-access slabs.

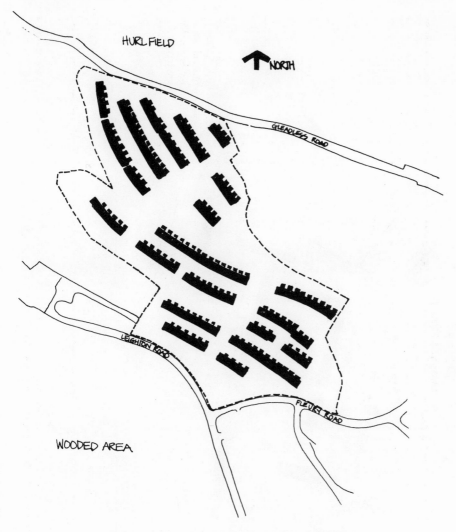

FIGURE 50. Site plan: Fleury Road, Sheffield.

Users

Personal background data from the interviews of 1314 housewives and 369 husbands provided the basis for defining four household categories which were then used to stratify the data for analysis purposes. The categories relate primarily to stages in the family life cycle.

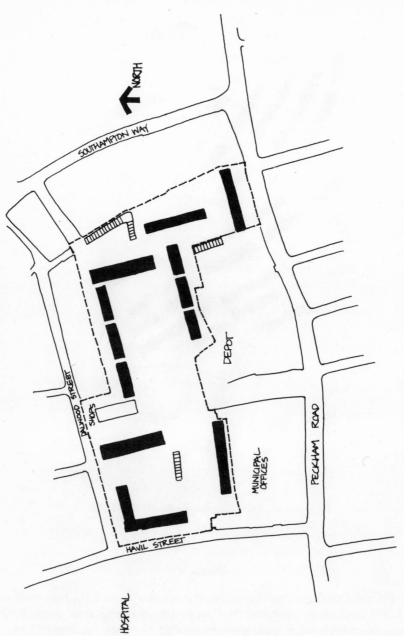

FIGURE 51. Site plan: Sceaux Gardens, Southwark.

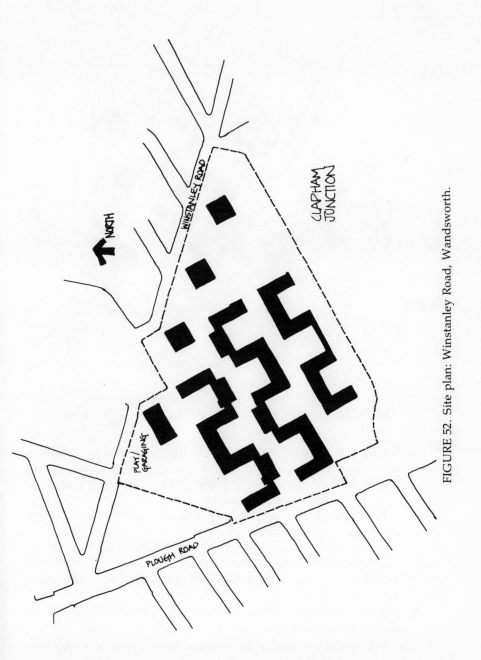

FIGURE 52. Site plan: Winstanley Road, Wandsworth.

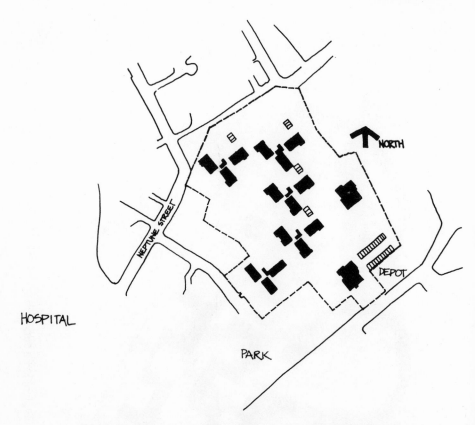

FIGURE 53. Site plan: Canada Estate, Greater London Council.

1. Family households—families made up of parents and children under 16: further subdivision of this category were families with (a) all children under 5, (b) some children under 5 and some over, and (c) all children 5 or over.
2. Adult households—all members 16 or over but no one of retirement age.
3. Mixed households—families with parents and children under 16 and with individuals of retirement age. (Because the samples of mixed households were quite small, they were not included in the final analysis.)
4. Elderly households—husbands and wives or single individuals all over retirement age.

Social class was defined by head-of-household occupation for family and adult households. Residents in all housing projects were pre-

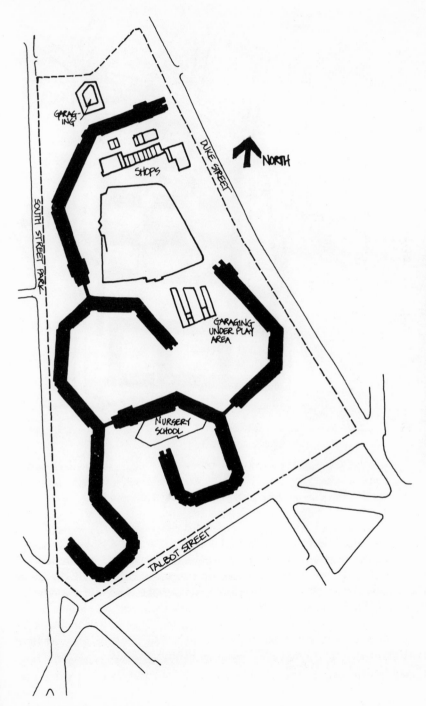

FIGURE 54. Site plan: Park Hill, Wheffield.

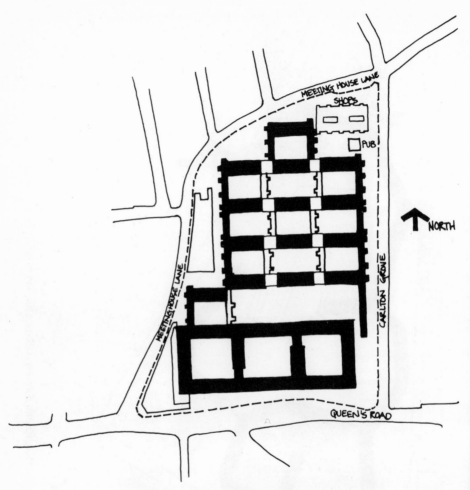

FIGURE 55. Site plan: Acorn Place, Southwark.

dominantly classified "skilled manual labor" (55 and 63%) and "semi-skilled" (21 and 22%) with a smaller percentage labeled "unskilled" (9 and 15%). Sixty percent of the family households, 74% of the adult households, and 100% of the elderly households had net weekly incomes in 1967 of less than 20 pounds (less than $50 per week in the U.S.).

Of the 1314 households in this study, 330 lived in ground-floor apartments or row houses and 984 lived off the ground.

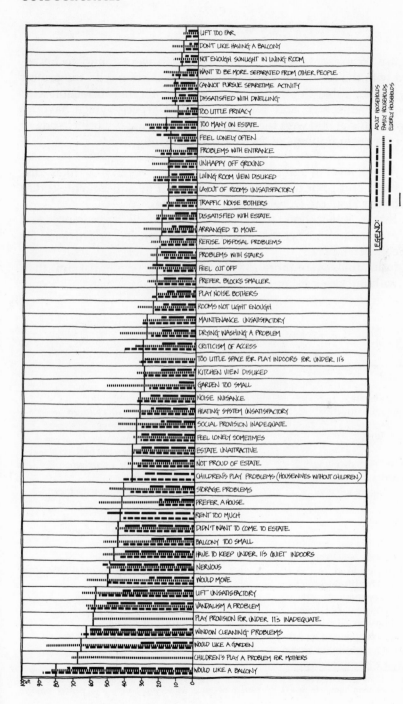

FIGURE 56. Responses to questionnaire indicating dissatisfaction with aspects of layout.

EVALUATION

Figure 56 presents those responses which were found to be associated with the residents' comparative dissatisfaction with the project (estate) layout, Table 25 presents those which indicate satisfaction with the estates themselves, and Table 26 presents aspects which have no bearing on satisfaction with the estates. The following is a brief listing of some of the main findings of the comparative study.

1. A high level of agreement among housewives in their likes, dislikes, and priorities (a finding borne out by subsequent studies by the Housing Development Directorate).
2. Estate satisfaction was related to surface appearance and maintenance rather than building form, density, or provision for children's play.
3. Individuals were more satisfied with their residential unit than with the larger estate outside of it.

TABLE 25
Aspects Related to Estate Satisfaction[a]

Closely related	Slightly related
.58 Proud of estate	.29 Play not a problem for housewives with children
.54 Appearance attractive	.28 Blocks not too large
	.28 Noise not a nuisance
Fairly closely related	.25 Play noise does not bother
	.25 Vandalism not a problem
.47 Satisfied with dwelling	.24 Not nervous
.43 Satisfied with maintenance	.23 Visitor's parking satisfactory
.43 Not arranging to move	
	.23 Access satisfactory
	.23 Wanted to come to estate
Related	.23 Not too little privacy
	.23 Heating satisfactory
.35 Living room view liked	.22 5–11 year olds play outside enough
.35 Kitchen view liked	.20 Sufficient space for play indoors
.33 Happy living off the ground	
.33 Prefer to be more separate	.20 Sufficient storage
.31 Like arrangement of rooms	
.31 Play provision satisfactory	
.30 Not too many people on estate	
.30 Play not a problem for housewives without children	

[a]Numbers indicate correlation between overall estate satisfaction and specific questionnaire responses. Possible correlation range from −1 (a perfect negative relationship) to +1 (a perfect positive relationship).

TABLE 26
Aspects Not Related to Estate Satisfaction

Socioeconomic characteristics	*Aspects of dwelling and estate*
Whether housewife works	Attitude toward traffic noise
Number of children	Whether present view important
Socioeconomic group of head	Whether have a garden
of household	Whether too much privacy
Social class of head of household	Sunlight
Possession of car	Daylight
Attitude toward rent	Window cleaning
Income	Size of balcony
Rent/income	Like balcony or not
	Want balcony if none
Housing background	Size of garden
	Want garden if none
Length of stay in present dwelling	Height of dwelling
Distance from previous dwelling	Drying/washing problems
Standard of previous dwelling	Refuse disposal problems
Length of stay in previous dwelling	Lived off ground previously or not
Satisfaction with previous dwelling	Change in attitude toward living
	off ground
Aspects of dwelling and estate	*Social aspects*
Use of access area for chatting	Nearness of close relatives
Use of access area for play	People known before coming to
Satisfaction with lift	estate.
Distance dwelling to lift	People met after coming to estate
Problems with stairs	Loneliness
Problems with entrances	Cut off
Whether activities restricted	Attendance at tenants' association
Whether sit out on estate	How often attend tenants'
	association
	Social provision on estate
Children's play	
	Priorities
Desire for under-fives to play	
outside more	Preference for spending more
Need to keep children quiet	on estate or dwelling
indoors	Preference for house on outskirts
Children safe from traffic	or flat in center of town

Regression analysis indicated that the best predictors of estate satisfaction and the percent of variance explained by each were: appearance (28%), dwelling satisfaction (10%), maintenance (8%), desire to move (2%), vandalism (1%), living room view (1%), and play problems (1%).*

*Percentages reflect the relative importance of the responses in predicting satisfaction. The percentages come from a statistical technique which supplements the correlations illustrated in Figure 25.

When all of the 47 other variables that accounted for 1% of the variability are considered, a total of 52% of the variance is explained.

DISCUSSION

A number of factors about this study merit special mention, including the selection of study sites, the number of interviews and the attention to physical as well as sociopsychological variables. The benefits to be realized from a comparative study are clearly illustrated in the findings concerning the relationship of such variables as building form and density to resident satisfaction. Studies that address only one building form or only one housing project do not provide as sound or reasonable a basis for generalizing to other situations; it is difficult, if not impossible, to assess whether effects can be attributed to specifics of the project or to more generic issues and problems.

The large interview sample enhanced the reliability of the findings. It also allowed for the disaggregation of the sample according to household type and thereby provided for a more detailed analysis of the data with potential implications for housing policy and design criteria. It provided, for example, the opportunity to question the suitability of various housing forms and locations for families at various stages in the life cycle.

The attention to detailed measurement of selected physical attributes of the projects such as balcony and garden size allowed for a more careful analysis of selected satisfaction ratings. The identification of relationships between satisfaction ratings and physical dimensions such as these provides an important basis for revising design guidelines and criteria.

Planned Unit Development Evaluation

Study by:	Michael Levin and Sandy Sachs of Rahenkamp, Sachs, Wells and Associates, Planners—Land Planners—Landscape Architects, Philadelphia
Methods used:	Interview, content analysis of rental applications
Type of project:	Resident evaluation of the Village of Pine Run, a planned unit development in New Jersey
Information source:	*People and Planning: Facts and Figures*, 1974, and correspondence with project designers

INTRODUCTION

The planned unit development (PUD) was considered a new and innovative land development concept in the 1960s. The concept encompasses a number of planning and design notions: cluster designs for housing and greater densities in developed areas, provision for a range of housing types, mixed residential, industrial, and commercial uses, separate treatment of pedestrian and vehicular circulation systems, analysis of the landscape to identify areas most suitable for development, and the preservation or protection of areas for important environmental processes or features. These notions also include a set of assumptions which is broadly shared by planners and designers about the qualities of the resultant environment and about users' perceptions of and attitudes toward such designed environments. This study reports on a user evaluation of a PUD in New Jersey, the Village of Pine Run (Figures 57 and 58).

The primary objective of the study was to assess planners' assumptions about user reactions to the PUD, assumptions such as:

1. A mixture of housing types with recreational and commercial uses provides a more balanced community (and implicitly, it would seem, a better community).

FIGURE 57. Site plan of Village of Pine Run, New Jersey.

2. Increased densities associated with the clustering of structures and with sufficient open space does not reduce the quality of life and is both more attractive and more marketable.
3. Preservation of open space is both a sound marketing and a sound ecological policy and contributes to amenity value.
4. Separate pedestrian and auto circulation systems are safer, more efficient, and more pleasant.

The evaluation was conducted by two planners who were staff members of the firm responsible for the planning of the PUD.

STUDY DESIGN AND METHODS

The primary data source was a structured interview administered to a stratified random sample of 80 households. The sample was stratified

FIGURE 58. Aerial view of Pine Run (courtesy of Rehenkamp, Sachs, Wells, and Associates).

so as to obtain proportional samples of the three housing types in the PUD: apartments ($N = 40$), townhouses ($N = 28$), and single-family detached dwellings ($N = 12$). The interviews were conducted during the summer of 1971.

The 70-question interview schedule was developed in consultation with sociologists from the University of Pennsylvania and consisted of six parts. Each part was oriented to a specific geographic or functional aspect of the PUD as follows:

1. The decision to move to Pine Run
2. The community of Pine Run (the overall community)

3. Courts/Sections (cluster or neighborhood in which the interviewee lived)
4. Specific residence (specific house, apartment, or townhouse)
5. Recreation activities
6. Travel

The design of the schedule included fixed-response questions, open-ended questions, and semantic-rating scales. Part 3, Courts/Sections, is illustrated in Figure 59.

Interview data were analyzed primarily in terms of percentage distribution of responses. These data were supplemented with personal background information which was obtained from rental application records.

Setting

Pine Run encompasses an area of 125 acres in Gloucester Township, New Jersey. At completion it will contain 713 residential units consisting of 74 single-family homes, 177 rental townhouses, and 462 rental garden apartments (see the site plan, Figure 57). The single-family homes and the townhouses are available at three different prices while the apartments are available at five different rental levels. Thus, potential residents have eleven alternatives to choose among. The PUD also includes recreational and community facilities, a shopping center, and a small industrial park. At the time of the study, Pine Run was about 80% complete and 500 of the residential units were occupied.

Prior to the study, a 4.5-acre convenience shopping center of some 17,500 square feet of floor space existed. However, it was nearly totally destroyed by fire. At the time of the study the only commercial facility remaining was a 7-Eleven store. A 3600 square foot community building and swimming pool were operational and adjacent to the 7-Eleven store. The 25-acre industrial park and a lake were in final planning stages.

Design Activity

The initial planning for Pine Run was developed in 1964; however, the PUD concept required zoning revisions which were not enacted until 1965. Late in the same year construction started.

The planning approach for Pine Run was predicated on the previously stated assumptions and started with an inventory of the natural features of the site (e.g., slopes, geology, soils, water table, and vegetation). This inventory identified which lands were most suited for development and which for preservation or protection as open space.

The PUD zoning regulations provided for the construction of 713

A

III Courts/Sections

In this section of the interview we are interested in your viewpoints concerning the particular court (homes) or section (apartments, townhouses) in which you live, in this case _____ .

---

(22) Defining "neighborhood" with the following characteristics (people at least knowing the names of other people, people saying hello to one another, perhaps borrowing things or doing favors once in a while), do you feel that a strong sense of neighborhood exists in your section or court?

a_____Strong sense of neighborhood   Explain:
b_____Some sense of neighborhood
c_____No sense of neighborhood
d_____Conflict
e_____No feelings on the subject

Using the following characteristics describe your particular court or section:

noisy	____	____	____	____	____ quiet
attractive	____	____	____	____	____ unattractive
unfriendly	____	____	____	____	____ friendly
poorly kept up	____	____	____	____	____ well kept up
crowded	____	____	____	____	____ uncrowded
similar people	____	____	____	____	____ different people
transient	____	____	____	____	____ permanent
liberal	____	____	____	____	____ conservative
adult oriented	____	____	____	____	____ child oriented
active	____	____	____	____	____ inactive
lack of privacy	____	____	____	____	____ private
others	____	____	____	____	____

Explain extreme responses:

With respect to the various characteristics just given, do you feel that there are significant differences between courts (or sections) within Pine Run:

_____

If yes, describe these differences (refer to map if necessary).

FIGURE 59. Sample portions of questionnaire.

mixed-type, clustered housing units on the most suitable land. Had the site been developed under the original zoning code, 360 single-family houses would have been allowed on 12,500 square foot lots.

The circulation system was designed to separate vehicles and pedestrians. The pedestrian walkways are located in the open spaces in

B

If you had the opportunity, would you move to another section/court:

_____

If yes, which one _____ and why:

_____

Have you made any complaints to the management:

_____

If yes, how many_____and what were they about:

_____

How responsive has the management been to these complaints:

a____Very responsive   Explain:
b____Responsive
c____Mixed feelings
d____Unresponsive
e____Very unresponsive

_____

(28)   Read the following 2 definitions of friendships.

       Neighbors: People you know by name, have occasional short con-
                   versations with, borrow things from, share car
                   pools, etc.

       Friends: People you see on a more regular basis and with whom
                   you do things socially (parties, visiting, going out,
                   etc.)

       Do you agree with these definitions: _____

       If no, give us your own definition of different friendship
       levels:

_____

Considering the above definitions (or your own) answer the following
two questions:

How often you socialize with your "neighbors" as compared with your
previous residence:

a____Socialize more   Explain (including differences
b____Same as before   among family members)
c____Socialize less
d____No thoughts on the subject

FIGURE 59. (continued)

C

How often you socialize with "friends" as compared with your previous residence:

a____Socialize more   Explain (including differences
b____Same as before   among family members):
c____Socialize less
d____No thoughts on the subject

---

(29) How do you feel about the number of residences located within
      your court/section:

      a____Very crowded   Explain:
      b____Somewhat crowded
      c____Adequate
      d____Not close enough

---

(30) Do you believe that there are better or worse locations within
      your section/court: _____

      If yes, describe and identify these differences:

---

(31) If you had the opportunity, would you move to another location
      within your section/court: _____

      If yes, where _____and why would you move there:

FIGURE 59. (continued)

the naturally maintained valleys. Roadway widths are scaled according to traffic movement and volume, and cul-de-sacs are used for single-family homes.

All of these design decisions relate to the initial assumptions about PUDs and residents' perceptions of the attendant environmental qualities.

Context

Pine Run is located about 30 miles south of the Philadelphia, Pennsylvania–Camden, New Jersey metropolitan area. It is bounded on the west by a multilane divided highway, the North–South Freeway. Development on the other three sides consists of rural–suburban scattered housing.

Residents/Users

Most of the data on the residents of Pine Run were obtained from the rental applications. These were supplemented with interview data for single-family home respondents.

Fifty-six percent of the residents had occupied rental housing prior to moving to Pine Run. Only 16% had been home owners. The balance had lived with parents. Seventy-nine percent were 35 years old or younger. The older residents tended to live in townhouses or single-family homes.

Twenty-four percent of the residents were single, living alone or with roommates; 45% were married without children and 31% were married with children. There were 449 children living in 37% of the total completed units. Thirty-three percent of the children were under 5 years of age, 56% were between 5 and 10, and the balance were over 18.

Forty-six percent of Pine Run residents were in education, professional, or managerial/executive occupations, 26% were in technical and sales work, and the remaining 28% were in clerical service, labor, and craft occupations. The majority of this last group were apartment dwellers.

The average annual income of the head of household was $13,188, although this varied considerably by housing type. The average annual income of heads of households for apartment residents was about $10,000, while the average for townhouse and single-family home heads of household approximated $17,000.

TABLE 27
What People Value in Pine Run (%)

Features	Adds a great deal	Adds somewhat	Neither adds nor detracts	Detracts somewhat	Detracts a great deal
Pool	51	34	10	4	1
Pedestrian paths	72	22	5	1	0
Wooded areas	94	4	0	2	0
Shopping facility	49	23	19	5	4
Roads	16	36	19	19	10
Parking areas	0	8	13	15	64
Community center	16	42	38	3	1

TABLE 28
Satisfaction with Community and Individual Residence (%)

	Extremely satisfied	Satisfied	Mixed feelings	Dissatisfied	Extremely dissatisfied
Community					
Apartment	10	33	48	7	2
Townhouse	17	54	25	4	0
Single family	0	84	16	0	0
Residence					
Apartment	26	40	22	5	7
Townhouse	33	38	25	4	0
Single family	8	50	33	8	0

EVALUATION

Table 27 presents summary responses for all study participants (apartment, townhouse, and single-family house) to questions dealing with features common to the PUD concept. The wooded open space areas are obviously universally viewed as an important asset, as are the pedestrian paths and the pool. Roads and parking areas are perceived less positively. This response may be, in part, attributable to the fact that only one parking space was provided per dwelling unit, thus causing some inconveniences for multicar families. Responses to detailed questions on roads indicated mixed perceptions as to safety and convenience but agreement on the overall attractiveness of the road system.

Responses to densities created by clustering were generally positive, with 74% rating density as adequate. Fifteen percent of the apartment residents, however, thought they were very crowded. Overall, 70% of the respondents also liked the idea of mixed housing.

Table 28 presents a comparison of satisfaction responses to the community and to individual residence units. In general, apartment and townhouse dwellers tend to be more satisfied with their residences than with the community. This is not the case for single-family home residents.

Table 29 indicates responses to evaluative questions about the individual residences. Visual concerns of attractiveness and privacy and the physical arrangement of rooms are perceived very favorably in all three residence types. A much more mixed response appears to questions regarding maintenance, the choice of materials, and heating and air conditioning.

TABLE 29
Residents' Evaluation of Individual Residential Unit (%)[a]

Scale item	Excellent			Good			Average			Below average			Poor		
	APT	TH	SF	APT	TH	SF	APT	TH	SF	APT	TH	SF	APT	TH	SF
Attractiveness of outside design	56	52	50	40	40	25	5	8	17	—	—	8	—	—	—
Attractiveness of inside	63	60	75	26	40	25	5	—	—	7	—	—	—	—	—
Room arrangement	50	48	75	45	48	25	5	4	—	—	—	—	—	—	—
Keeping out noise	33	48	25	38	40	42	20	12	17	5	—	17	5	—	—
Visual privacy	40	52	50	44	36	33	9	12	17	2	—	—	5	—	—
Outside open space	49	44	50	26	32	25	16	16	25	5	8	—	5	—	—
Kitchen facilities	35	32	33	47	56	33	7	12	17	9	—	8	2	—	8
Sunlight	21	36	58	42	48	33	16	12	9	7	4	—	14	—	—
Maintenance	10	19	—	31	52	18	29	16	27	14	8	36	17	4	18
Landscaping	40	28	—	43	30	29	10	16	25	2	16	33	5	8	33
Colors	33	36	50	48	48	50	14	16	—	2	—	—	2	—	—
Materials	5	8	8	14	17	33	16	42	8	33	17	43	33	17	8
Heating/air conditioning	14	20	0	23	40	33	7	24	33	30	12	16	26	4	16
How much are you getting for your money?	23	25	25	50	50	58	23	25	17	3	—	—	3	—	—

[a] APT, apartment; TH, townhouse; SF, single-family detached dwelling.

DISCUSSION

This study represents an approach to evaluation that should be within the means of any competent professional office of medium to large size. The data-gathering and analysis techniques are simple, direct, and informative. However, the data are also suitable for more sophisticated and rigorous statistical analysis. The cost to the office in the early 1970s was approximately $15,000.

The use of social science consultants in the development of the interview schedule is an important element in protecting against biases that might otherwise be inadvertently designed into the instrument. The mix of question formats in the interview schedule (fixed-response and open-ended questions and rating scales) should help to maintain interviewee interest. Questions requiring the comparative ranking of community, neighborhood, or residence attributes and features might have been a desirable addition to provide checks on the reliability of responses to questions dealing with independent ratings of attributes or features such as indicated in Tables 28 and 29.

The use of nonobtrusive methods such as the analysis of rental records is another important and useful attribute of the study. It provides a basis for cross-tabulation of the data according to selected socioeconomic characteristics of residents as well as by type of residence. It is also a more effective and impersonal way of obtaining information that might be given rather reluctantly in an interview setting.

In summary, the study tended to confirm many, but not all, of the initial planning assumptions. It provided guidance to the professionals as to which attributes should be emphasized and which should be reconsidered. For example, the notions about the desirability of having a socially heterogeneous proximate population (which was facilitated by alternative designs and rental rates) was not strongly shared by the residents. The Rahenkamp Office reports that they are now experimenting with designs which are supportive of more homogeneous clusters or neighborhoods. Also, the office no longer advocates the expense of community centers in similar developments. The study report is used as a way of introducing new employees to the concern for client/user-oriented design solutions.

V

Environmental Design Evaluation: Epilogue

Several pieces of the jigsaw puzzle of environmental design evaluation have now been presented; this chapter will fill in the remaining pieces. First, the direction of the field will be charted, both generally and by the use of specific cases. Second, following the conclusion that better methodology is needed, several broad issues will be presented, including both general issues of validity and summaries of techniques. However, these techniques involve some important ethical dangers; these are outlined in the fourth section. Finally, the structure–process approach will be summarized in the last section.

FUTURE DIRECTIONS

We have presented an overview of environmental design evaluation and have discussed several case studies. Yet where is the field headed? What goals should be set? Although there is no consensus on these issues, several themes have been discussed in this book and elsewhere.

(1) It is critical that we increase the influence of evaluation in the design–evaluation–design cycle. There are a number of ways to do so. (a) Evaluation should be taught to both design students and working professionals. (b) The academic reward system must be changed so that evaluation activities are given due academic recognition alongside the more conventional research and publication activities. (c) More attention needs to be given to the style and format of reports, especially those directed at designers. This might include redoubled attempts to use clear, nonjargonistic prose and more extensive use of graphics.

(2) Evaluations need to be financially feasible for designers. Often even highly successful firms do not have the time or money to complete evaluations. In publicly funded construction there needs to be contractual provisions similar to program evaluation components of federal

social programs. For example, after a design contract is let it may be increased by 5% for evaluation. Like federal programs, funding of design projects would be contingent on an articulated evaluation component. The focal problem, larger system, methods, and personnel would have to be spelled out in detail. Clients in the private sector might follow suit, especially those clients who often repeat similar types of projects such as apartments, housing, institutions, schools, or offices.

(3) There is too much separation between application and theory. Direct, physically oriented evaluations are often carried out by designers; theoretically oriented evaluations are carried out by social scientists. Yet, theory and application are synergistic. Every new design contains new and unique elements: a theory of what people need in their environment; what they can comprehend; what is beautiful to them; what makes them happy. These are critical questions for behavioral sciences, too. By better integrating theory and application, design evaluation can refine theory which, in turn, improves subsequent applications.

(4) Data banks of evaluation information need to be established. This pool of information could be coordinated jointly by a consortium of design schools and the relevant professional organizations and could serve both educational and professional purposes. Like the Medical Literature Analysis and Retrieval System, MEDLARS, the data bank can be computer-coded, with information retrieved through key words describing the information factors of the structure–process (i.e., users, settings, proximate environmental context, design activity, social–historical context).

However, if such a data bank is to be meaningful, there needs to be better research agenda. Professional organizations should address the question: "What is the highest priority for evaluation?" A research agenda might not only address the question of priority of settings but also help to ensure that evaluations occur in a broad sampling of settings from both the public and private sectors.

Finally, a data bank requires better, more standardized methods. Although each setting is different and requires a somewhat unique approach, the use of common questionnaires, observation schemes, and interview schedules aids comparison of data from different settings. This standardization might come in part as books such as this one are written, making common techniques more readily available. In part, also, standardization becomes easier if there is a better understanding of methodology, of what techniques are appropriate for specific situations.

METHODOLOGY, TOOLS, AND TECHNIQUES

Success or failure of an evaluation often depends on the skill with which an evaluator selects and uses information-gathering techniques. These methods should be simple, clear, straightforward, and should efficiently gather information needed. In research, as in many endeavors, simplicity often requires the greatest skill.

A complete discussion of methods is beyond the scope of this book. Methodological discussions fill hundreds of volumes; indeed, many people spend their entire professional careers refining a single method such as questionnaire or interview. In this section we will quickly sketch out some broad methodological issues, then devote a few paragraphs to several specific techniques. This is not a "how to do it" chapter. It is, rather, a brief summary of some of the methods used in the case studies. The reader is referred to one of the excellent books listed in the bibliography for more complete help.* This discussion will serve as an introduction for readers unfamiliar with the relevant issues and will be a springboard into more detailed presentations.

It is important to understand the general characteristics of environmental design evaluation research. Rather than *manipulating* environments the way experimental studies do, evaluation usually seeks to *describe* what is going on. Because they do not artificially induce activities, these descriptive techniques should give a clearer understanding of activities which would occur without the intervention of the researcher. Despite this asset, however, the lack of control attendant to descriptive methodology is sometimes maddening when the evaluator's plans must change due to an unforeseen change in the settings. Similarly, the most powerful conceptual and statistical tools of social research are often inappropriate because of the lack of control that the evaluator has: rather than searching for the *cause* of behavior, the evaluator must be reconciled to understanding a few of many influences, at best. These differences between evaluation and experimental research underlie the following comments.

Several issues are central to most aspects of social research. These include such questions as: Who should be sampled and when? Are the measuring techniques sufficiently sensitive to record the users' activities or reactions which form the focal problem? Is the designed setting actu-

---

*Brandt (1972) and Michelson (1975) are useful summaries of methods. Other more in-depth sources are included in the bibliography.

ally causing the measured activities (or, are there other causes present, such as the presence of the evaluator)? To which other settings and groups do the findings generalize? How much consistency is present in the data? Do different evaluators record data similarly? Do different data-gathering methods agree? Does activity change with time and/or setting?

Multiple information-gathering methods are used in most evaluation research. This strategy of using converging techniques allows the weakness of one method to be partially compensated by the strength of another. For example, coding observed activities into predefined categories may allow quite sensitive measurement. This is due to the volume of data generated and the interobserver consistency which can be achieved. However, unexpected activities may not be accurately recorded because no category has been previously established. Interviews may be able to record such changes but may allow less sensitive measurement. No matter how many methods are used, however, it is critical to plot out the analysis *before the data gathering* commences. Whether the analytic methods are statistical or not, prior planning helps prevent the collection of a lot of uninterpretable "garbage." It is also highly desirable for whatever methods are decided upon to go through pretesting prior to the adaption of questionnaires or other formalized procedures. A few sample interviews, a few observations, or a few completed questionnaires will reveal potential problems and enable the researcher to modify them before going ahead with the larger study.

Information-gathering methods can be described in five categories: *direct observation*, techniques in which the user's activities are directly recorded; *interview*, where users are asked their reactions to settings; *unobtrusive measures*, indirect measures for tapping user activities; *simulation*, methods where users react to artificial representation of environments; and *pencil-and-paper* tests, written instruments to understand user activities.

CRITICAL ISSUES IN EVALUATION

Sampling

This topic is often discussed in terms of four issues. First, a general distinction is often made in field research between "opportunity sampling" and "random sampling." In the former case, participants are chosen as opportunity permits. For example, managers may be asked to furnish names of volunteers who will fill out a questionnaire. Opportunity sampling presents a clear possibility of bias. The sample may not represent the larger group: managers may chose people who are favorably disposed toward the settings, who are attractive, or are good workers. In random sampling the evaluator chooses participants on a chance basis, perhaps drawing names out of a hat. Randomization increases representativeness of the sample, yet may be impractical in some situations. For example, it may be possible to randomly choose every fifth customer entering the store, yet relations with the store management may preclude random sampling of employees. In general, random sampling is preferred. Many studies do not allow it, however, and opportunity sampling is quite acceptable.*

A second type of randomization is even more rarely available: random assignment of users to settings. Ideally, we would like to know how different types of people reach to a housing design, office arrangement, or other settings. Yet, in most cases people select their environments: some people choose houses, others apartments. So we often only know how a certain type of user reacts to a setting and cannot generalize to the larger population.

A third sampling issue is choosing the appropriate user group. This was discussed in some depth in Chapter I. It suffices at this point to emphasize that all groups affected by a design need to be considered "users" and need to be sampled accordingly.

Finally, the sampling of settings is a critical concern of environmental design evaluation. It was suggested in Chapter I that several types of settings are of high priority for evaluation, such as those which affect the public or are publicly financed. However, the choice of settings affects generalizability just as does the choice of participants. Comparative studies of several settings extend data and help clarify how generally applicable those findings are, whereas single-site evaluations may not do so. Also, choosing sites which are representative of a larger class of settings increases generalizability as does choosing diverse *types* of settings.

Reliability

If evaluation data are to be useful, they must be consistent in at least two ways. First, there must be some assurance that the way activities are

---

*In large-scale survey research random sampling often occurs in a framework where user types are differentially chosen to represent their occurrence in the larger population. Such stratified sampling is best left to an outside consultant or other expert.

recorded is not entirely dependent on the particular evaluator doing the recording. The recording should reflect a commonly agreed upon interpretation of the ongoing activity, not one idiosyncratic to an individual. To check this agreement, interrater reliability is often computed in research using direct observation. Two observers simultaneously code the same behavior and their agreement is tabulated. An agreement of 70–90% usually serves as the minimum criterion to be achieved before observation begins, depending on the complexity of the observation scheme.

Second, the data should show some stability across time. As we have seen in Chapter I, activities change as society evolves over time. However, if evaluation data are to be useful, they must be consistent for time spans of a week, months, or even a few years. Feedback into the design process requires that some elements will remain stable through the next iteration of the design.

The stability depends in part on the methods used and in part on the activities observed. If activities are observed only on a single day, there is a good chance that the data will be unstable—the one particular day chosen might be special. However, more stable patterns often emerge if observations occur over weeks or months. Also, some user responses such as attitudes toward management may be more variable from day-to-day than are others such as basic feelings about the importance of "home."

Validity

Issues of validity reflect the quality of an evaluation. For example, *internal validity* refers to sensitivity of the data-gathering instruments to detect changes in focal issues. For example, in a before-and-after questionnaire study of renovations, are the questionnaires sensitive enough to detect differences in attitude between the two environments? Internal validity is generally increased when more participants are used and when other confusing influences (the larger system) are carefully documented and understood. (In experimental social science, internal validity is increased by using standardized laboratory conditions. This is seldom possible in environmental design evaluation.)

Construct validity generally refers to the adequacy with which the causes or effects of the influences are identified. For example, a study with good internal validity may detect a change; but is it really due to the renovations? Perhaps the evaluator became more familiar with the participants and recorded their responses differently, or other outside influences such as management changes affected the attitudes. There is no way to guarantee construct validity; however, using reliable instruments and documenting the larger system help considerably in understanding true influences.

A third type of validity, *external validity*, is concerned with the generalizability of the information. This was briefly mentioned in the section on sampling. To what other situations do the findings generalize? In environmental design evaluation, both the physical setting and the users need to be considered. For example, an evaluation of a carpeted open-plan school could be expected to generalize to other identical schools, yet what about uncarpeted schools? Or to semiopen plans? Such generalizations are, of course, qualified by the users. An evaluation of an open-plan school for white middle-class users may be quite different from an evaluation of a similar school for students from other cultural traditions. External validity is directly related to the accuracy of the description of setting, users, context, and so on. If these factors are well described, it becomes much easier to understand which environments are similar.

Specific Methods

Table 30 summarizes the different methods employed in the case studies. The methods are further explained in the following sections.

Direct Observation

A commonly used technique in environmental design evaluation, direct observation of activity may be divided into two categories according to method: narrative or checklist (Brandt, 1972).

Narrative. According to Brandt, the narrative type includes data that attempt to reproduce behavioral activities in much the same fashion and sequence as they originally occurred. Rather than pigeonholing activities in explicit, rigid categories, the observer attempts to record ongoing events as he/she sees them. For example, the evaluator may record incidents in detail where people are interacting with specified aspects of the setting. Analysis of such anecdotes may highlight strengths and weaknesses of the design. Alternatively, the evaluator may try to record *all* activity in a given space or by given people. This is often called a specimen record. Specimen records provide a large amount of information and their all-inclusive nature usually results in a high degree of agreement among observers. Finally, the use of field notes is the most common narrative technique. Field notes combine the two previous tech-

TABLE 30

Methods Used in Case Studies Reviewed in Chapters II, III, IV

	Direct observation	Interview	Unobtrusive measures	Simulation	Pencil-and-paper tests
Chapter II: Interior Spaces					
Office landscape		Open-ended and focused			Questionnaires and open-ended
Butterfield Hall	Open-ended, nonparticipant				Questionnaires and open-ended URES (University Residence Environment Scale)
Subway station	Tracking (following selected individuals) Behavior mapping (location of whole group)				
Cambridge Hospital	Photography Open-ended, nonparticipant	Focused			Questionnaire
Fine Arts Center		Focused	Archival records about design process		Focused

Chapter III:
Buildings-as-Systems

ELEMR Project	Behavior mapping, time sampling, participant observation, open-ended nonparticipant observation, photography	Open-ended, focused, critical incidents	Analysis of institutional tests and records		Critical incidents
New England Villages	Open-ended, participant and non-participant, photography		Wear measures, analysis of records of design process		
Charlesview Housing	Open-ended, photography	Focused, open-ended	Analysis of records of design process		
Visitor centers	Tracking, behavior mapping	Open-ended, focused	Analysis of records of design process	Photographs of alternative sites	Questionnaires (open-ended and focused)

Chapter IV:
Outdoor Spaces

Campus space	Tracking of users				Newspaper questionnaire

(continued)

TABLE 30 (continued)

	Direct observation	Interview	Unobtrusive measures	Simulation	Pencil-and-paper tests
Chapter IV: Outdoor Spaces					
First National Bank Plaza	Photography, open-ended nonpartici-pant	Open-ended	Analysis of documents about design process		
Bryant Park	Behavior mapping, participant	Open-ended and focused	Analysis of documents about design process		Focused questionnaire
Housing site	Open-ended	Focused	Measurement of physical environment		
Planned Unit Development (PUD)		Focused	Analysis of rental applications		

niques, allowing the evaluator to select specific anecdotes at some times and to record all behaviors at others.

Because the evaluator using narrative techniques imposes relatively little structure on data, they allow for the recording of unexpected patterns of activities that may emerge. On the other hand, if the evaluator using anecdotal methods freely selects incidents to be recorded, that evaluator's bias obviously influences the choice and recording of anecdotes. Common measures for reducing bias include having prearranged criteria for picking incidents (e.g., all incidents of a specific type, every fifth conversation), recording incidents promptly, using direct quotes wherever possible, and using physical rather than evaluative descriptions (e.g., "red" rather than "beautiful"). Specimen records are said to be more objective than anecdotal ones because all behavior is recorded—no choice of incidents is presumably required. Each observer, however, interprets his or her world in a slightly different manner, and hence even specimen records vary. Objectivity may be increased by keeping the description of activity in fairly simple terms and separating direct recording from theorizing and conjecture. Specimen records produce a large quantity of information. It is often hard to reduce these records to a manageable size.

Checklists. Brandt has suggested: "Whenever the existence or nonexistence of specific objects, conditions, or events needs to be recorded systematically and consistently, the checklist is a promising device" (1972, p. 94). One common use of checklists is to record the physical facilities themselves. Items such as maintenance conditions of various areas, barrier-free access, life-safety issues, and amount of decoration can all be contained in a checklist. Also, user activity is increasingly being recorded by checklists of various sorts. Two common types of activity checklists are used: event sampling and time sampling. In event sampling, specific, well-defined events are recorded (e.g., two clerks in a department talking). When the specific defined events do not occur, no mark is made. Time sampling generally employs an observation scheme with several mutually exclusive and exhaustive categories. Any possible activity can be coded into one (but only one) of the prescribed categories. Observations occur at prearranged times; there is one mark for each time. For example, behavior mapping is a commonly used time-sampling technique. At prearranged times an observer codes the activities and locations of all the people in a space.

An advantage of checklists is that coding is relatively quick and simple once the categories are decided. Also, if category definitions are sufficiently clear and unambiguous, it is quite easy to achieve consensus among observers. However, establishing such categories is often quite difficult. Also, checklists may undervalue important but infrequent acts (e.g., the visit of the company president) and may not allow the coding of unexpected activities where categories are not preestablished.

Interview

Perhaps the most commonly used tool for assessing people's reactions to designed settings, Interviews can be divided into two types: structured (or formal) and unstructured. Structured interviews are those in which the type and order of questions are decided in advance. These interviews range from Harris-poll-type questions where respondents are read questions verbatim, to checklists where key words are listed for each question and the wording of the question is left to the discretion of the interviewer. The first method yields results which are comparable among respondents. Yet because it is not flexible, the questions may be inappropriately stated for some respondents. The checklist-type technique allows for greater naturalness, but because questions may be phrased differently, responses may be more difficult to compare.

Unstructured interviews take place in the course of normal conversation. The interviewer slips in a few questions of interest while visiting the site or waiting in a reception area. In this manner the interviewer gains information which is natural and unbiased and which is produced in the "natural habitat" of the informant. The interviewer knows the questions that he or she wants to ask, yet can feel free to alter their order, or to explore some in more depth than others as the natural sequence of the conversation allows.

Interviewing has several important advantages for the evaluator. Because it takes advantage of the knowledge of people already involved in the setting, it provides a fast way to gain understanding. The interview is adaptable to fit different needs and different groups. For example, unstructured interviews may be appropriate while establishing basic issues, while more structured approaches might be better for actually addressing the focal problem. Also, interviewing provides very rich information. Feelings, motives, anecdotes can all be gleaned by interview but are hard to measure by direct observation or other methods.

There are also drawbacks with interviews. Interviews are time-consuming and hence expensive. Since much skill is required to get relatively "unbiased" results, even the findings of professional interviewers are often in question. Furthermore, the responses to questions are often unstructured verbal responses which are hard to quantify and compare. Finally, in many situations the respondents feel that their job (and/or their privacy) are in jeopardy and may produce stereotyped or expected responses. It may take some time to create the rapport necessary to gather honest answers.

Unobtrusive Measures

A serious problem exists with most information-gathering techniques: they influence activities they are intended to measure. The presence of an observer or camera may cause people to behave in an unnatural manner; an interviewer may give unintentional cues about the answer that he or she expects to receive. Several years ago, Webb *et al.* (1966) presented a number of *post hoc* techniques that do not affect the ongoing behaviors.

Document Review. In almost any evaluation, many documents are available, and, in general, researching these documents does not affect the activities in the setting. For example, the design process may be researched by reading correspondence and design programs and by examining working drawings. Institutional records show employee absenteeism rates, requests for transfer, production figures, and so on. Such records may also document maintenance costs and other problems with the physical settings. Also, a great many public documents exist which may illuminate different facets of the evaluation, such as magazines, newspapers, *Moody's Handbook, Who's Who in America*, tax records, and so on.

Advantages of record searching are that it generally does not affect the ongoing activities, provides alternative measures, and can give insight into activities at earlier times (e.g., can reveal the architect's original intentions). However, records are often kept haphazardly. The accuracy and completeness of documents may depend on particular recordkeepers and may vary with time or department. For this reason it may be hard to see how activities have changed across time or between settings. Good, reliable records are usually available; however, care must be taken in their use.

Physical Traces. Often the way a setting is used can be seen by physical traces. For example, the tiles in front of popular exhibits in a museum may be particularly worn or may require frequent replacement. Similarly, the height of noseprints on the glass of a display case may indicate the relative age of viewers. Broken windows and litter may show lack of pride in facilities. The number and types of locks on doors may show the amount of fear about crime. The evaluator uses these measures much the way an archeologist does; by capitalizing on the various marks people leave on their environment.

An advantage of measuring physical traces is that, like using documents, the act of measuring usually does not affect ongoing activity. Unlike direct observation, worn tiles and broken glass can be measured when no one except the evaluator is present. Also, these unobtrusive measures are independent of other observation and interview techniques and can corroborate those methods. Finally, such measures speak to issues of maintenance and cleanliness, which are high priority for managers and administrators.

The problems with using physical traces usually relegate them to the role of secondary, corroborative methods. Primarily, it is difficult to know the precise *cause* of most traces. Worn tiles may indeed indicate heavy use of an exhibit but may also indicate turnings in a hallway or nearness to a washroom or to a drinking fountain. Litter may show lack of pride or heavy use of an area.

Simulation

Loosely speaking, in the "simulation" method respondents' comments are evoked from representations of settings rather than from the settings themselves. Computer graphics, videotape, photographs, verbal descriptions, drawings, and models are all simulations. Indeed, simulation may be the most underused method in evaluation.

There are many potential uses of simulation. It is often cheaper and quicker to have people respond to simulations than to the settings themselves, especially when the settings are remote or widely scattered. Also, simulations such as computer graphics can be quickly altered, allowing users to manipulate designs in a way that is not possible with real environments.

However, there are drawbacks to simulation. Users are not reacting to the whole experience of the setting; they cannot understand the subtle successes and failures that emerge only when living with a built design. Some research has shown, however, that photographs and drawings can reproduce visual experience, and that more elaborate simulations such as videotapes can give even more lifelike experiences.

Pencil-and-Paper Tests

Most users of designed environments are literate and used to dealing with written questions; questionnaires and other written instruments may be useful and efficient methods to survey these users.

Questionnaires. Questionnaires use several different information-gathering formats. The semantic differential is frequently used to understand both explicit denotive aspects of environments (e.g., large, colorful) and more subjective connotative aspects (e.g., joyful, restful). Opposite terms (hot–cold) are usually displayed together and the responder has to mark some point on a continuum between them. Although this adjective-pair format has become familiar to many users and hence is useful, the use of the semantic differential in environmental research has caused some recent criticism: The semantic differential was developed to explore meaning structures of abstract concepts (such as "church"), and it is not clear how the differential operates when used to assess actual environments. Also, the connotative and denotative meanings of environments are quite complex. Even with language, the exact distribution of meanings affects the analysis of results, and language is probably less complex than are many-faceted environmental experiences (Bechtel, 1975).

Other common questionnaire items include both open-ended and fixed-response questions, and, in general, questionnaires use both types of items. Open-ended questions allow the respondent to compose his or her own answer to questions such as "What do you like most about this setting?" The utility of such an approach is that the evaluator does not impose his or her preconceived notions of appropriate responses on the respondent. Generally, some sort of content analysis is used in analyzing open-ended responses.

Fixed-response questions have prearranged response formats such as a multiple-choice or a multipoint continuum between two labeled points (e.g., agree–disagree). Some other graphic formats are also quite well developed (Bechtel, 1975). The strength of fixed-response formats is that they allow relatively easy comparison between respondents and facilitate statistical manipulation. However, by providing specific responses, the evaluator may suggest outcomes that otherwise might not have occurred to the respondent. Also, of course, the range of outcomes must be clear to the evaluator beforehand. For this reason fixed-response format may be of greatest use late in an evaluation when the specific focal issues are clearest.

In general, questionnaires are useful and efficient data-gathering methods. However, special care must be devoted to determining whether the respondents understand the questionnaire and are accustomed to working in a written mode. This is not always obvious! (For example, we found that a number of staff members at an institution were uncomfortable understanding written instructions and had to be interviewed.) Also, the order of questions may affect results. Earlier questions may suggest the thrust of the inquiry and influence later answers. A final serious concern is that some answers may appear more socially desirable than others and may be picked for that reason, rather than because they reflect "true" attitudes. Because of these many considerations, it is often desirable for the inexperienced researcher to have help from an experienced person, especially for longer and more complete questionnaires.

Other Pencil-and-Paper Tests. Although the questionnaire is by far the most common written instrument, many others are in use. For example, researchers using "cognitive mapping" ask respondents to describe in some way their conception of the environment. Most researchers ask people to sketch their environment on a blank sheet of paper (Lynch, 1960); computers and other technologies are now being used which allow for more complex analysis of cognitive maps. Although cognitive mapping has primarily been used to assess city- or neighborhood-scale issues, it holds promise for site- and single-building-scale research. Mapping can shed light on such diverse topics as wayfinding within buildings and sites, the relative importance of different areas within a setting, and the effectiveness of signage systems. Methodological problems (such as the scoring of cognitive maps) remain, however (Moore and Golledge, 1976). Other pencil-and-paper tests which are not as well developed are described by Bechtel (1975).

ETHICAL CONCERNS

Any research endeavor entails unique ethical and moral problems. Environmental design evaluation is especially prone to controversy because of the variety of interests that must be reconciled. There are several critically important ethical issues which need to be especially considered in evaluation: (1) respect for the participants and (2) political concerns and conflicts of interest.

Respect for Participants

The specific concerns under this principle flow from a general concern for people who are serving as informants or are otherwise involved in an evaluation. For example, one overriding concern is to obtain consent of participants. This involves taking time to describe the evaluation in at least a general way and calming fears about participating (e.g., "We're looking over this building and I'd like to ask you a few questions about it. Of course, we'll never reveal your name."). Moreover, recent government guidelines have required written "informal consent." For interviews, it is usually taken as sufficient consent if a participant is fully told about the uses of the data and then verbally agrees to participate.

More elaborate written consent forms are usually required if participants are observed or filmed.

Interviews, observation, or examination of records often uncover sensitive information, and care must be taken in using this information. For example, some people hold unorthodox political views, have aspects of their past history they want to conceal, or are concerned about their job performance. Confidences must be respected: when the evaluator gains trust of the participant and gains confidential information, he or she must take absolute measures to protect the privacy of the participants. For example, names of participants should not be recorded or should be immediately coded into numbers; tapes or transcriptions of interviews should be kept secure and not shown to management or the press; observation data should be coded and made anonymous.

Also, the personal privacy of the participants must be respected, especially because evaluations often study private spaces such as bedrooms, bathrooms, and private offices. Some methods such as direct observation or filming may not be appropriate for private spaces, whereas they might be for public areas. And, of course, the evaluator must attempt to minimize the impact of data-gathering itself for research reasons as well as ethical ones. The generalizability of information may be reduced if the study itself has a large effect on participants. Assessing these impacts often requires considerable sensitivity on the part of the evaluator and constant communication with the participants.

In addition, respect for participants carries with it respect for professional and personal relationships that have been made. It may take days or weeks to establish one's self in the setting, to establish rapport, and identify informants. The relationships made during this process should be respected both for the evaluation at hand and for future evaluations. For example, a summary of results of the evaluation should be given to the participants, appointments and other commitments should be honored, and a fair chance should be given for participants to react to results.

Political Concerns and Conflicts of Interest

Most evaluations face conflicting pressures about the definition of the problem and form of the results. For example, evaluators often have conflicting loyalties. The client for the evaluation is often a design firm, building owner, or other group with a vested interest in the outcome of the evaluation, and there may be pressure toward positive findings. (This pressure may be even greater if a firm evaluates its own designs.) However, the evaluator is part of a larger professional research community which has high standards of quality and objectivity. In addition, there are a broad range of users that must be considered: tenants, custodial workers, managers, office workers, and so on. Each of these groups wants its concerns addressed in an evaluation.

There is no single answer to these many conflicts. However, some general considerations may be of help. One approach suggested by J. Reizenstein (1977, personal communication) is that the threatening nature of evaluations is often overly stressed and unnecessarily creates conflict:

> The reason for conducting POEs (postoccupancy evaluations) is to learn, not to judge, and the reason for learning about how environments accommodate the various needs of their users is to use that information to make future environments better. Unfortunately, the more threatening approach of judging environmental designers, owners, or managers may be implied in POE. One reason for this may be that environment–behavior shortcomings ("misfits") may be more obvious than fits.

In other words, some conflicts surrounding an evaluation may be lessened by treating it more as a learning experience and less as a way to assign blame. Social science is by nature critical and often searches for aspects of situations that do not work or should be changed. This is a valid scientific stance, yet in evaluation it is equally important to document what *is* working and what *should* be kept.

There are several approaches to conflict of interest. For example, whereas there are good reasons for firms to evaluate their own designs, this situation also brings problems of bias. The problem may be partially offset by having one or more outside consultants on the evaluation team. Also, in such situations it is desirable to use methods such as observation into coded categories which are less susceptible to bias in addition to using open-ended interviews or participant observation.

Most basically, these conflicts take on different significance depending on the eventual use of the evaluation. An in-house evaluation which is intended to fine-tune an existing design may be legitimately less concerned with a comparative sample of users from different projects than would an evaluation which is intended to generalize more broadly.

STRUCTURE–PROCESS APPROACH

Environmental design evaluation consists of studies with a common objective: assessing the effectiveness of designed environments for users. These studies have several goals, such as fine-tuning existing environments, providing information for new designs, and providing basic information about human activity in designed settings. Although it is possible to spend much time and money on an evaluation, a short and

Summary

inexpensive study may be quite productive if it is carefully thought out and well designed.

A two-faceted structure–process approach was presented in Chapter 1: (1) a five-part conceptual *structure* for organizing information and (2) a five-step *process* for actually accomplishing evaluations.

The structure allows the evaluator to place large volumes of necessary information in a useful framework. The five factors which comprise the structure include: (1) *setting*—insight into social and physical attributes of the designed project being evaluated; (2) *users*—an understanding of background, needs and activities of people who are involved with the setting, such as tenants, customers, maintenance workers, managers, and passersby; (3) *proximate environmental context*—an understanding of ambient air and water qualities, land-use characteristics, and neighborhood qualities that surround the setting; (4) *design activity*—insight into activities by designers, regulatory agencies, and clients which resulted in the final design of the setting; and (5) recognition that the previous four factors exist in a larger *social–historical context*, the society in which one must consider social, economic, and policy issues such as social mores, unemployment levels, and demographic profiles.

After needed information is organized, the evaluation must be performed. As described in Chapter I, a multistep process for completing the evaluation must be employed. First, several preevaluation issues must be resolved. For example, the responsibility for initiating and compiling design-oriented evaluation rests with the design professionals themselves. Courses must be taught and data banks of information created. Also, evaluation teams should be composed of both scientists and designers, although there are potential problems with such collaboration. Finally, the choice of the setting to be evaluated must be made, and there are a number of settings which are of high priority for evaluation. These include settings of broad generalizability or theoretical importance.

The evaluation process itself includes several steps: defining the *focal problem* (relationships of central interest in the evaluation); defining the *larger system* (other influences of importance); defining methods and gathering data; and analyzing data.

The entire jigsaw puzzle has been presented. However, it is unlike the familiar picture-puzzle where the player merely has to match up a bit of horse or sky or trees to create a simple, familiar scene. The "puzzle" that is environmental design evaluation has many solutions, many "scenes" of equal interest and importance. It is left to the reader to assemble the puzzle in a unique and individual way.

Selected Bibliography

This bibliography serves three purposes: (1) it provides bibliographic information for works referenced in the text; (2) it serves as a research bibliography as it includes selected additional works of interest; (3) it provides complete bibliographic information for cases discussed in Chapters II, III, and IV. It is based on the *Environmental Design Evaluation Bibliography*, by E. Zube, C. Zimring, and J. Crystal (Amherst, Ma.: Institute for Man and Environment, 1975).

I. A CONCEPTUAL APPROACH TO ENVIRONMENTAL DESIGN EVALUATION

Altman, I. Some perspectives on the study of man–environment phenomena. *Representative Research in Social Psychology*, 4:(1973), 109–126.

American Society of Civil Engineers. *A Positive Interprofessional Approach to the Environmental Impact Statement. Sponsored by the Interprofessional Council on Environmental Design.* Summary and recommendations of the conference at Arlie House, Arlie, Va., Nov. 28–30, 1972.

American Society of Landscape Architects. Priorities For ASLA. *ASLA Bulletin* (February, 1974).

Appleyard, D. *BART Traveler Environment: Environmental Assessment Methods for Stations, Lines and Equipment.* Berkeley: University of California, Institute of Urban and Regional Development, May, 1973.

Appleyard, D. The Second International Architectural Psychology Conference at Lund, Sweden. *Landscape Architecture 64*:(1973), 53–54.

Appleyard, D. *Environmental Planning and Social Science: Strategies for Environmental Decision Making.* Working Paper No. 217. Berkeley: University of California Institute of Urban and Regional Development, 1973.

Appleyard, D. *Liveable Urban Streets: Managing Auto Traffic in Neighborhoods.* Washington D.C., Superintendent of Documents, U.S. Government Printing Office, 1976.

Appleyard, D. and F.M. Carp. The BART residential impact study: An empirical study of environmental impact. In *Environmental Impact Assessment: Guidelines and Commentary,* (ed. T.G. Dickert and K.R. Domeny). Berkeley: University of California Institute of Urban and Regional Development, 1974.

Appleyard, D., *et al. The Berkeley Environmental Simulation Laboratory: Its Use in Environmental*

Impact Assessment. Berkeley: University of California Institute of Urban and Regional Development, 1973.

Baker, M.S. Implication of the National Environmental Policy Act. In *Environmental Design Research*, Vol. 2. (ed. W.F.E. Preiser). Stroudsburg, Pa.: Dowden, Hutchinson, and Ross, 1973.

Bouterline, S. The concept of environmental management. In *Environmental Psychology* (ed. H.M. Proshansky *et al.*). New York: Holt, Rinehart and Winston, 1970.

Braybrooke, S. Evaluating evaluation. *Design & Environment* 5(3):(Fall, 1974), 20–25.

Brill, M. Evaluating buildings on a performance basis. In *Designing for Human Behavior* (ed. J. Lang *et al.*). Stroudsburg, Pa.: Dowden, Hutchinson and Ross, 1974.

Brolin, B.C. Chandigarh was planned by experts but something has gone wrong. *The Smithsonian* (June, 1972).

Brolin, B.C., and J. Zeisel. Mass housing: Social research and design. *The Architectural Forum* (July/August, 1968), 66–71.

Cambell, D. Evaluation of the built environment: Lessons from program evaluation. In *The Behavioral Basis of Design, Book 1: Selected Papers* (ed. P. Suedfeld and J. Russell). Stroudsburg, Pa.: Dowden, Hutchinson and Ross, 1976.

Carp, F., and D. Appleyard. *Pre-BART Traveler Attitudes and Perceptives: East Bay Panel, BART Impact Studies Part II*. Vol. 1. Berkeley: University of California, Institute of Urban and Regional Development, May, 1973.

Conway, D. (ed.). *Social Science and Design*, Washington, D.C.: American Institute of Architects, 1973.

Cooper, C.C. *Easter Hill Village*. New York: The Free Press, 1975.

Cooper, C., and P. Hackett. *Analysis of the Design Process at Two Moderate-Income Housing Developments*. Berkeley: University of California, Institute of Urban and Regional Studies, 1968.

Council on Environmental Quality, *Environmental Quality*, The Eighth Annual Report of the Council, Washington D.C.: U.S. Government Printing Office, 1977.

Craik, K.H., and E.H. Zube. *Perceiving Environmental Quality, Research and Application*. New York: Plenum Press, 1976.

Davis, T.A. Evaluating for Environmental Measures. In *Proceedings of the 2nd Annual Environmental Design Research Association Conference*, 1970, 45–55.

Eckbo, G. Evaluating the Evaluation. *Design & Environment:* 2(4): (Winter, 1971), 39–40.

Festinger, L., S. Schachter, and K. Back. *Social Pressure in Informal Groups*. Stanford, Ca.: Stanford University Press, 1963.

Gutman, R. Site Planning and Social Behavior. *Journal of Social Issues* 22 (4): (Oct., 1966), 103–115.

Gutman, R. *People and Buildings*, New York: Basic Books, 1972.

Gutman, R., and B. Westergaard. Building Evaluation, User Satisfaction and Design. In *Designing for Human Behavior* (ed. J. Lang, C. Burnette, W. Moleski, and D. Vachon). Stroudsburg, Pa.: Dowden, Hutchinson and Ross, 1974.

Hollingshead, A.B., and L.H. Rogler. Attitudes towards slums and public housing in Puerto Rico. In *Human Behavior and the Environment* (ed. J.H. Sims and D.D. Baumann). Chicago: Maaroufa Press, 1974.

Kaplan, R. Preference and everyday nature: Method and application. In *Perspectives on Environment and Behavior: Theory, Research, and Applications* (ed. D. Stokols). New York: Plenum Press, 1977.

Kaplan, R. Participation in the design process: A cognitive approach. In *Perspectives on Environment and Behavior: Theory, Research, and Applications* (ed. D. Stokols). New York: Plenum Press, 1977.

Kaplan, R. *Environmental Design: The Participation Model*. (mimeo), Department of Psychology, University of Michigan, Ann Arbor, 1973.

Katz, R.D., and D.G. Saile. *Activities and Attitudes of Public Housing Residents: Rockford, Illinois.* Urbana-Champaign: University of Illinois, Committee on Housing Research and Development, 1970.

Knight, R.C., Zimring, C.M., and M.J. Kent. Normalization as a social–physical system. In *Barrier-Free Environments* (ed. M.J. Bednar). Stroudsburg, Pa, 1977.

Kurtz, S.A. and L. Fink. And now a word from the users. *Design & Environment* 3 (1):(Spring, 1971), 41–48.

Lang, J., and W. Moleski. A Behavioral theory of design? *Design and Environment* 4:(1973), 44–47.

Lang, J., C. Burnette, W. Moleski, and D. Vaihon. *Designing for Human Behavior.* Stroudsburg, Pa.: Dowden, Hutchinson and Ross, 1974.

Lansing, J.B., and R.W. Marans. Evaluation of neighborhood quality. *Journal of the American Institute of Planners* 35:(May, 1969), 195–199.

Lansing, J.B., R.W. Marans, and R.B. Zahner. *Planned Residential Environments.* Ann Arbor, Michigan: University of Michigan, Institute for Social Research, 1970.

Lynch, K. *The Image of the City.* Cambridge: MIT Press, 1960.

McHarg, I. *Design With Nature.* New York: Natural History Press, 1969.

Michelson, W. *Man and His Urban Environment: A Sociological Approach,* 2nd ed. Reading, Mass.: Addison-Wesley, 1976.

Newman, O. *Defensible Space.* New York: Macmillan, 1972.

Norcross, C. *Townhouses and Condominiums: Residents' Likes and Dislikes.* Washington, D.C.: Urban Land Institute, 1973.

Ostrander, E. The visual–semantic communication gaps. *Man–Environment Systems* (1974), 47–53.

Ostrander, E., and B.R. Cornell. Maximizing cost–benefits of post-construction evaluation. In *The Behavioral Basis of Design, Book 1: Selected Papers* (ed. P. Suedfeld and J. Russell). Stroudsburg, Pa.: Dowden, Hutchinson and Ross, 1976.

Ostrander, E., and J. Reizenstein. *Creative Living: Housing for the Severely Disabled in the Context of a Service Delivery System.* Washington,. D.C.: American Institute of Architects Research Corporation, 1976.

Perin, C. *With Man in Mind: An Interdisciplinary Prospectus for Environmental Design.* Cambridge, Mass.: MIT Press, 1970.

Porteous, J.D. *Environment & Behavior: Planning and Everyday Urban Life.* Reading, Mass.: Addison-Wesley, 1977.

Preiser, W. *Behavioral Design Criteria. Student Housing, Research Report No. 1.* Blacksburg, Va.: Virginia Polytechnic Institute, College of Architecture, Environmental Systems Labs, 1969.

Rabinowitz, H.A. *'Buildings In Use' Study.* Milwaukee: University of Wisconsin, School of Architecture, 1975.

Rainwater, L. Fear and the house-as-haven in the lower class. *Journal of the American Institute of Planners* 32:(1966), 23–31.

Rapoport, A. The design professions and the behavioral sciences. *Architectural Association Quarterly* 1(1):(1969), 20–24.

Rapoport, A., and W.F.E. Preiser (ed.). *Environmental Design Research,* Vol. 2. Stroudsburg, Pa.: Dowden, Hutchinson, and Ross, 1973.

Reizenstein, J. Linking social research and design. *Journal of Architectural Research* 6(1):(March, 1977).

Sanoff, H., S. Christie, D. Tester, and B. Vaupel. Building evaluation. *Building International* 6:(1973), 261–297.

Sommer, R. *Personal Space: The Behavioral Basis of Design.* Englewood Cliffs, N.J.: Prentice-Hall, 1969.

Sommer, R. *Design Awareness.* San Francisco: Rinehart Press, 1972.

Sommer, R. Evaluation, yes: Research, maybe. *Representative Research in Social Psychology* 4(1):(January, 1973), 127–134.

Stokols, D. Experience of crowding in primary and secondary environments. *Environment and Behavior 8*:(1976), 49–87.

Weddle, A.E. *Techniques of Landscape Architecture*. New York: American Elsevier, 1967.

Wheeler, L. Behavioral Research for Architectural Planning and Design. Terre Haute, Indiana: Archonics Corp., 1967.

Whyte, W.H. In *A Study of the Profession of Landscape Architecture* (ed. A. Fein). Washington, D.C.: American Society of Landscape Architects Foundation, 1972a.

Whyte, W.H. Please, just a nice place to sit. *N.Y. Times Sunday Magazine* (Dec. 3, 1972b), pp. 20ff.

Wohlwill, J.F. The environment is not in the head! In *Environmental Design Research*, Vol. 2 (ed. W.F.E. Prieser). Stroudsburg, Pa.: Dowden, Hutchinson and Ross, 1973.

Zeisel, J. Symbolic meaning of space and the physical dimension of social relations: a case study of sociological research as the basis of architectural planning. In *Cities in Change: Studies on the Urban Condition* (ed. J. Walton and D. Carns). Boston: Alyn and Bacon, 1973.

Zeisel, J. *Sociology and Architectural Design*. New York: Social Science Frontiers: Occasional Publications Reviewing New Fields for Social Science Development, Russell Sage Foundation, 1974.

Zube, E. H. *A Multi-Factor Approach to Site Design Evaluation*. Amherst, Ma: University of Massachusetts, Institute for Man and Environment, 1974.

Zube, M. *Community of Peers: Private and Public Lives of Residents in Housing for the Elderly*. Dissertation (unpublished). Amherst, Ma.: University of Massachusetts, 1974.

II. INTERIORS

Case Studies

Brookes, M. and A. Kaplan. Office environments: Space planning and effective behavior. *Human Factors 14*:(1972), 373–391.

Brown, M., T. Johnson, Y. Kishimoto, L. Reynolds, S. Suzuki, and K. Tepel. *An Evaluation of the Fine Arts Center at the University of Massachusetts, Amherst*. Amherst, Ma.: University of Massachusetts, 1975 (unpublished).

REDE (Research and Design Institute). *Butterfield Hall Evaluation Report*. Providence, R.I.: REDE, 1974.

Reizenstein, J.E., K.R. Spencer, W.A. McBride. *Social Research and Design: Cambridge Hospital Social Services Offices* (unpublished manuscript). Cambridge: Harvard University, Graduate School of Design, 1976.

Winkel, G.H., and D.G. Hayward. *Some Major Causes of Congestion in Subway Stations*. New York: Environmental Psychology Program, Graduate Center, City University of New York, 1971.

Items Cited and Works of Interest

Boyce, P.R. Users' assessment of a landscaped office. *Journal of Architectural Research* 3(3):(September, 1974), 45–63.

Hayward, G. Research techniques in an evaluation of subway station designs. *DMG–DSR Journal* 6(1):(1972), 17–19.

Lau, J.J.H. Differences between full-size and scale model rooms in the assessment of lighting quality. In *Architectural Psychology*, (Proceedings of the Conferences held at Daladhue, University of Stathclyde, Feb. 28–March 2, 1969) (ed. D. Carter). London: RIBA Publications Limited, 1970.

Manning, P. Office design: A study of environment. In *Environmental Psychology* (ed. H.M. Proshansky *et al.*) New York: Holt, Rinehart and Winston, 1970.

Moleski, W. Behavioral analysis and environmental programming for offices. In *Designing for Human Behavior* (ed. J. Lang, C. Burnette, W. Moleski, and D. Vachon). Stroudsburg, Pa.: Dowden, Hutchinson and Ross, 1974.

Osmond, H. Function as the basis of psychiatric ward design. In *Environmental Psychology* (ed. H.M. Proshansky *et al.*). New York: Holt, Rinehart and Winston, 1970.

Probst, R.L., and C.G. Probst. *The University of Massachusetts Dormitory Experiment.* Ann Arbor, Mi.: Herman Miller Research Corp., 1973.

Stephens, S. AIA headquarters: Magnificent intentions. *Architectural Forum* (October, 1973).

III. BUILDINGS-AS-SYSTEMS

Case Studies

Knight, R.C., C.M. Zimring, W.H. Weitzer, and H. Wheeler. *Social Development and Normalized Institutional Settings: A Preliminary Research Report.* Amherst, Ma.: University of Massachusetts, Environment and Behavior Research Center, Institute for Man and Environment, 1977.

Reizenstein, J., and W. McBride. *Designing for Mentally Retarded People: A Social–Environmental Evaluation of New England Villages.* Ann Arbor, Mi.: University of Michigan, Architectural Research Laboratory, forthcoming.

Zeisel, J., and M. Griffin. *Charlesview Housing: A Diagnostic Evaluation.* Cambridge, Ma.: Architecture Research Office Graduate School of Design, 1975.

Zube, E.H., J.H. Crystal, and J.F. Palmer. *Visitor Center Design Evaluation.* Amherst, Ma.: University of Massachusetts, Institute for Man and Environment, 1976.

Items Cited and Works of Interest

Adelberg, T.Z., and M.W. Shelley. Notes on Satisfactions in Shopping Center, I, II, III, and IV. *Psychological Reports* 21:(October, 1967), 507–508, 536, 584, 660.

Bechtel, R.B. The public housing environment: A few surprises. In *Proceedings of the 3rd Annual Environmental Design Research Association Conference,* 1972.

Boudon, P. *Lived-in Architecture: Le Corbusier's "Pessac" Revisited.* Cambridge, Ma.: MIT Press, 1969.

Cooper, C. *Easter Hill Village.* New York: Free Press, 1975.

Cooper, C. St. Francis Square: Attitudes of its residents. *American Institute of Architects Journal* 56(6):(December, 1971), 22–27.

Cooper, C., and P. Hackett. *Analysis of the Design Process at Two Moderate Income Housing*

Developments. Working Paper No. 80. Berkeley: University of California, Institute of Urban and Regional Development, 1972.

Cooper, C., N. Day, and B. Levine. *Resident Dissatisfaction in Multi-Family Housing.* Working Paper No. 160. Berkeley: University of California, Institute of Urban and Regional Development, 1972.

Deasy, N., and T. Lasswell. *Real Goals Versus Popular Stereotypes in Planning for a Black Community: The Hooper Avenue School Study.* New York: Educational Facilities Laboratories, 1966.

Griffin, M.E. Mount Hope Courts: A social–physical evaluation. In *Proceedings Northeastern Undergraduate Conference, Environment and Behavior* (ed. J. Vogt). Amherst, Ma.: University of Massachusetts, Institute for Man and Environment, 1974, 33–49.

Ministry of Housing and Local Government (U.K.). Old people's flatlets at Stevenage. In *People and Buildings* (ed. R. Gutman). New York: Basic Books, 1972.

Sauer, L., and D. Marshall. How six families use space in their existing houses. In *Proceedings of the 3rd Annual Environmental Design Research Association Conference,* 1972.

Tica, P., and J.A. Shaw. *Barrier-Free Design, Accessibility for the Handicapped.* New York: City University of New York, Center for Advanced Study in Education, Institute for Research and Development in Occupational Education, 1974.

Van der Ryn, S., and H. Silverstein. *Dorms at Berkeley: An Environmental Analysis.* New York: Educational Facilities Laboratories, 1967.

Wolfensberger, W. The normalization principle and some implications to architectural–environmental design. In *Barrier-Free Environments* (ed. M.J. Bednar) Stroudsburg, Pa.: Dowden, Hutchinson and Ross, 1977.

Zeisel, J., and B.C. Brolin. Mass housing: Social research and design. *Architectural Forum* (July/August, 1968). 66–71.

Zeisel, J., and M. Griffin. Charlesview Housing: A diagnostic evaluation. In *Proceedings Northeastern Undergraduate Conference, Environment and Behavior* (ed. J. Vogt). Amherst, Ma.: University of Massachusetts, Institute for Man and Environment, 1974, 19–22.

IV. BUILDING COMPLEXES AND OUTDOOR SPACES

Case Studies

Cohen, H., J. Crystal, J. Pflager, R. Rosenthal, and H. Wheeler. *Design Evaluation of a Central Outdoor Space at the University of Massachusetts.* Amherst, Mass.: University of Massachusetts Student Center for Educational Research, 1976, p. 72.

Levin, M., and S. Sachs. *People and Planning: Facts and Figures.* Philadelphia, Pa.: Rohenkamp, Sachs, Wells and Associates (1717 Spring Garden St.), 1974, p. 222.

Nager, A.R., and W.R. Wentworth. *Bryant Park: A Comprehensive Evaluation of its Image and Use with Implications for Urban Open Space Design.* New York: City University of New York, Center for Environment and Behavior Studies, 1976.

Reynolds, I., and C. Nicholson. *The Estate Outside the Dwelling, Design Bulletin 25.* London: Her Majesty's Stationery Office, 1972.

Rutledge, A.J. *First National Bank Plaza, Chicago, Ill. A Pilot Study in Post Construction Evaluation.* Urbana, Ill.: University of Illinois, Department of Landscape Architecture, June, 1975, p. 64.

Items Cited and Works of Interest

Bechtel, R. *Arrowhead: Final Recommendations*. Kansas City, Mo.: Environmental Research and Development Foundation, 1971.

Becker, F.D. Evaluating the Sacramento Mall. *Design & Environment 2*(4):(Winter, 1971), 38.

Becker, F.D. A class-conscious evaluation: Going back to Sacramento's pedestrian mall. *Landscape Architecture 64:* (1973), 448–457.

Cook, J.A. Gardens on housing estates: A survey of user attitudes and behavior on seven layouts. *Town Planning Review 39:*(1968), 217–234.

Deasy, C.M. People watching with a purpose. *American Institute of Architects Journal 54*(6):(December, 1970), 35–40.

Ertel, M. *Evaluation and Programming of Urban Teenage Hanging Places*. Cambridge, Mass.: Harvard Graduate School of Design, Architecture Research Office, 1974.

Grey, A., D. Bonsteal, A. Winkel, and R. Parker. *People and Downtown Use, Attitudes, Settings*. Seattle: University of Washington, College of Architecture and Urban Planning, September, 1970, 150.

Hayward, D.G., M. Rothenberg, and R.R. Beasley. Children's play and urban playground environments: A comparison of traditional, contemporary and adventure playground types. *Environment and Behavior 6*(2):(June, 1974), 131–168.

Hester, R.T., Jr., C. Long, D. Palmer, and E. Schweitzer. *User Needs as Design Determinants. Eight Case Studies*. Raleigh: North Carolina State University, School of Design, 1974.

Hole, W.V., and A. Miller. Children's play on housing estates. *Architects Journal 143:*(June 22, 1966), 1529–1533, 1535–1536.

Kaiser, E.J., et al. *Neighborhood Environment and Residential Satisfaction: A Survey of the Occupants and Neighborhoods of 166 Single Family Homes in Greensboro, North Carolina*. Chapel Hill, N.C.: University of North Carolina at Chapel Hill, Center for Urban and Regional Studies, October 25, 1970.

Lansing, J.B., R.W. Marans, and R.B. Zehner. *Planned Residential Environments*. Ann Arbor: University of Michigan, Survey Research Center, 1970.

Lerup, L. Environmental and behavioral congruence as a measure of goodness in public space: the case of stockholm. *DMG–DRS Journal 6*(2):(1972), 54–78.

Lyle, J.T. People watching in parks. *Landscape Architecture 61*(1):(1970), 31, 51–52.

Marcus, C.C. Children in residential areas: Guidelines for designers. *Landscape Architecture 64:*(1974), 372–377, 415–416.

Norcross, C. *Townhouses or Condominiums: Residents' Likes and Dislikes*. Washington, D.C.: Urban Land Institute, 1973.

Preiser, W.F.E. Analysis of pedestrian velocity and stationary behavior in a shopping mall. *Man–Environment Systems 4:*(1974), 63–64.

Rahenkamp, Sachs, Wells, and Associates. *Pine Run*. Philadelphia: 1974.

Saile, D. *Activities and Attitudes of Public Housing Residents: Rockford, Illinois*. Urbana-Champaign: University of Illinois, Department of Architecture, 1971.

Saile, D. *Families in Public Housing: An Evaluation of Three Residential Environments in Rockford, Illinois*. Urbana-Champaign: University of Illinois, Department of Architecture, 1972.

Sommer, R. Behavioral evaluation of a bikeway system. In *Proceedings Northeastern Undergraduate Conference, Environment and Behavior* (ed. J. Vogt). Amherst, Ma.: University of Massachusetts, Institute for Man and Environment, 1974, pp. 12–18.

Sommer, D., and F.D. Becker. The old man in Plaza Park. *Landscape Architecture* 60(1):(January, 1969), 111–113.

White, L.E. The outdoor play of children living in flats: an inquiry into the use of court-yards and playgrounds. In *Environmental Psychology* (ed. H.M. Proshansky *et al.*). New York: Holt, Rinehart and Winston, 1970.

V. ENVIRONMENTAL DESIGN EVALUATION: EPILOGUE

General

Becker, H.S. Whose side are we on? *Social Problems 14:*(Winter, 1967), 239–247.

Brandt, R.M. *Studying Behavior in Natural Settings.* New York: Holt Reinhart and Winston, 1972.

Braybrooke, S. Watching a people watcher. *Design and Environment 5*(3) (Fall, 1974), 26–29.

Brill, M. Evaluating buildings on a performance basis. In *Designing for Human Behavior* (ed. J. Lang, C. Burnette, W. Moleski, and C. Vachon). Stroudsburg, Pa.: Dowden, Hutchinson and Ross, 1974.

Campbell, D.T., and J.C. Stanley. *Experimental and Quasi-Experimental Designs for Research.* Chicago: Rand McNally, 1963.

Chapin, S.F., Jr., and T.H. Logan. Patterns of time and space use. In *The Quality of the Urban Environment* (ed. H.S. Perloff). Baltimore, Md.: The Johns Hopkins Press, 1969.

Dean, J.P., and W.F. Whyte. How do you know if the informant is telling the truth? *Human Organization 17*(2):(1958), 34–38.

Goodey, B. Displays for mating. *Design and Environment 3:*(1972), 46–53.

Harmatz, M.G. Observational study of ward staff behavior. *Exceptional Children 39:*(April, 1973), 554–558.

Kaplan, A. *The Conduct of Inquiry: Methodology for Behavioral Science.* Scranton, Pa.: Chandler, 1964.

Lang, J., C. Burnette, W. Moleski, and D. Vachon (eds.). *Designing for Human Behavior.* Stroudsburg, Pa.: Dowden, Hutchinson and Ross, 1974.

Lindberg, G., and H. Hellberg. Strategic decisions in research design. In *Behavioral Research Methods in Environmental Design* (ed. W. Michelson). Stroudsburg, Pa.: Dowden, Hutchinson and Ross, 1975.

Lindzey, G., and E. Aronson. *The Handbook of Social Psychology, Vol. II, Research Methods.* Reading, Ma.: Addison-Wesley, 1968.

Michelson, W. (ed.). *Behavioral Research Methods in Environmental Design.* Stroudsburg, Pa.: Dowden, Hutchinson and Ross, 1975.

Moos, R.H. *Evaluating Treatment Environments.* New York: Wiley, 1974.

Moos, R.H. *Evaluating Correctional and Community Settings.* New York: Wiley, 1975.

Rainwater, L., and D.J. Pittman. Ethical problems in studying a politically sensitive and deviant community. *Social Problems 14:*(Spring, 1967), 357–366.

Sanoff, H. Measuring attributes of the visual environment. In *Designing for Human Behavior* (ed. J. Lang *et al.*) Stroudsburg, Pa.: Dowden, Hutchinson and Ross, 1974.

Sommer, R. Some costs and pitfalls in field research. *Social Problems 19:*(Fall, 1971), 162–166.

Sommer, R. The new evaluator cookbook. *Design and Environment,* 2(4) (Winter, 1971), 24–35.

Thiel, P. Notes on the description, scaling, notation, and scoring of some perceptual and cognitive attributes of the physical environment. In *Environmental Psychology* (ed. H. Proshansky *et al.*). New York: Holt, Rinehart and Winston, 1970.

Wax, R. *Doing Fieldwork*. Chicago: The University of Chicago Press, 1971.

Techniques

As, D. Observing environmental behavior: The behavior setting. In *Behavioral Research Methods in Environmental Design* (ed. W. Michelson). Stroudsburg, Pa.: Dowden, Hutchinson and Ross, 1975.

Bechtel, R.B. The semantic differential and other paper- and pencil tests. In *Behavioral Research Methods in Environmental Design* (ed. W. Michelson). Stroudsburg, Pa.: Dowden, Hutchinson and Ross, 1975.

Becker, H.S. Problems of inference and proof in participant observation. *American Sociological Review*, Vol. 23. 1958, pp. 652–660.

Becker, H.S., and B. Greer. Participant observation and interviewing: a comparison and comments, In *Issues in Participant Observation* (ed. G. McCall and J. L. Simmons). Reading, Mass.: Addison-Wesley, 1969.

Bruyn, S.T. *The Human Perspective in Sociology: The Methodology of Participant Observation*. Englewood Cliffs, N.J.: Prentice-Hall, 1966.

Cannell, C.F. and R.L. Kahn. Interviewing. In *The Handbook of Social Psychology*, Vol. II, *Research Methods*, 2nd ed. (ed. G. Lindzey and E. Aronson). Reading, Mass.: Addison-Wesley, 1968.

Collier, J., Jr. *Visual Anthropology: Photography as a Research Method*. New York: Holt, Rinehart and Winston, 1967.

Davis, G. and V. Ayers. Photographic recording of environmental behavior. In *Behavioral Research Methods in Environmental Design* (ed. W. Michelson). Stroudsburg, Pa.: Dowden, Hutchinson and Ross, 1975.

De Jonge, D. Applied hodology. *Landscape* 17(2):(1967), 10–11.

Finstenbusch, K., and C.P. Wolf (eds.). *Methodology of Social Impact Assessment*. Stroudsburg, Pa.: Dowden, Hutchinson and Ross, 1977.

Godschalk, D. Negotiate: An experimental planning game. In *EDRAI* (ed. H. Sanoff and D. Cohn). Raleigh: North Carolina State University, 1970.

Holsti, O.R. Content analysis. In *The Handbook of Social Psychology*, Vol. II, *Research Methods* (ed. D. Lindzey and E. Aronson). Reading, Mass.: Addison-Wesley, 1968.

Lynch, K. *Image of the City*, Cambridge, Ma.: MIT Press, 1960.

Marons, R. Survey research. In *Behavioral Research Methods in Environmental Design* (ed. W. Michelson). Stroudsburg, Pa.: Dowden, Hutchinson and Ross, 1975.

McCall, G.J. Data quality control in participant observation. In *Issues in Participant Observation* (ed. G. McCall and J.L. Simmons), Reading, Ma.: Addison-Wesley, 1969.

McCall, G.J., and J.L. Simmons. *Issues in Participant Observation*. Reading, Ma.: Addison-Wesley, 1969.

Medley, D.M., and H.E. Mitzel. Measuring classroom behavior by systematic observation. In *Handbook of Research on Teaching* (ed. N. Gage). Chicago, Ill.: Rand McNally, 1963.

Merton, R.K., M. Fiske, and P.L. Kendall. *The Focused Interview: A Manual of Problems and Procedures*. New York: Free Press, 1956.

Moore, G., and R. G. Golledge. *Environmental Knowing*. Stroudsburg, Pa.: Dowden, Hutchinson and Ross, 1975.

Sanoff, H., and G. Coates. Behavioral mapping: An ecological analysis of activities in a residential setting. *International Journal of Environmental Studies* 2:(1971), 227–235.

Schulberg, L. Behavior mapping for design. *Design and Environment* 2(1):(Spring, 1971), 34–35.

Vidich, A.J., and G. Shapior. A comparison of participant observation and survey data. In *Issues in Participant Observation* (ed. G. McCall and J. L. Simmons). Reading, Ma.: Addison-Wesley, 1969.

Webb, E.J., D.T. Campbell, R.D. Schwartz, and L. Sechrest. *Unobtrusive Measures: Nonreactive Research in the Social Sciences.* Chicago: Rand McNally, 1966.

Weick, K.E. Systematic observational methods. In *The Handbook of Social Psychology*, Vol. II, *Research Methods* (ed. G. Lindzey and E. Aronson). 2nd ed. Reading, Ma.: Addison-Wesley, 1968.

Wells, B. Individual differences in environmental response. In *Environmental Psychology* (ed. H.M. Proshansky *et al.*). New York: Holt, Rinehart and Winston, 1970.

Whyte, W.F. Interviewing in Field Research. In *Human Organization Research* (ed. R.N. Adams and J.J. Preiss). Homewood, Ill.: Dorsey Press, 1960.

Analysis

Archea, J., and C. Eastman (eds.) *Proceedings of the Second Annual Conference of the Environmental Design Research Association.* Pittsburgh, Pa.: University of Pittsburgh Press, 1970.

Linton, M., and P.S. Gallo, Jr. *The Practical Statistician: Simplified Handbook of Statistics.* Monterey, Calif.: Brooks/Cole, 1975.

Lofland, J. *Analyzing Social Settings: A Guide to Qualitative Observation and Analysis.* Belmont, Calif.: Wadsworth, 1971.

Mitchell, J. (ed.) *Environmental Design Research and Practice; Proceedings of the Third Annual Conferences.* Los Angeles: University of California Press, 1972.

Mosteller, F., and J.W. Tukey. Data analysis, including statistics. In *The Handbook of Social Psychology*, Vol. II, *Research Methods.* (ed. G. Lindzey and E. Aronson). 2nd ed. Reading, Ma.: Addison-Wesley, 1968.

Index